Michael Maestlin's Manuscript Treatise on the Comet of 1618

Medieval and Early Modern Philosophy and Science

VOLUME 33

The titles published in this series are listed at *brill.com/memps*

Michael Maestlin's Manuscript Treatise on the Comet of 1618

An Edition and Translation of Manuscript WLB Stuttgart, Cod. Math. 4° 15b, Nr. 8

Edited by

Miguel Á. Granada
Patrick J. Boner

BRILL

LEIDEN | BOSTON

Cover illustration: The path of the third comet of 1618, from Wilhelm Schickard, *Cometen Beschreibung In zwen underschidliche Partes abgetheilt*, Württembergische Landesbibliothek Stuttgart, Cod. math. qt. 43, f. 86r /old Pag. 157. Also included as Figure 5.

The Library of Congress Cataloging-in-Publication Data is available online at http://catalog.loc.gov/
LC record available at http://lccn.loc.gov/2022008863

Typeface for the Latin, Greek, and Cyrillic scripts: "Brill". See and download: brill.com/brill-typeface.

ISSN 2468-6808
ISBN 978-90-04-47219-8 (hardback)
ISBN 978-90-04-51264-1 (e-book)

Contents

Preface

Michael Maestlin (1550–1631), Professor of Mathematics and Astronomy at the University of Tübingen and one of the few adherents to Copernicanism in the period between Copernicus and Kepler-Galileo, was struck—like every astronomer of his time—by the celestial novelties that appeared in the heavens from 1572 onwards. His treatises in Latin on the nova of 1572 in Cassiopeia and on the comets of 1577 and 1580 count among the most significant contributions on these phenomena. His study of the comet of 1577, in particular, marks for many the dawn of modern cometary theory. Maestlin pioneered a more complete mathematical analysis of comets, tracing their paths and positions beyond the sublunary sphere.[1] His Copernicanism was manifest from his tracts on the nova and the comet of 1577, and it was duly noted and criticised by Tycho Brahe.[2] The Prince of Astronomers praised Maestlin for his accuracy and insight,[3] but ultimately disagreed with his cosmological conclusions.

Maestlin was also attentive to the ensuing celestial novelties and wrote two manuscript treatises on the nova of 1604 in Serpentarius and the comet of 1618. These treatises, now among Maestlin's papers at the Württembergische Landesbibliothek in Stuttgart (WLB in what follows), were never printed, although their existence was always known to scholars, at least since Richard A. Jarrell wrote the first modern intellectual biography of Maestlin in his doctoral dissertation of 1971.[4] These treatises—the first on the nova written in Latin and the second on the comet of 1618 in German—remained unfinished, in draft form, with many deletions in the main text, additions in the margin (also frequently deleted), and in a hand that is sometimes very difficult to decipher. Jarrell's descriptions of the manuscript on the nova as "badly scored and illegible" and of the first two chapters of the one on the comet as "badly scored and practically illegible",[5] may have contributed to the fact that both treatises have been largely neglected by scholars.

Recently, however, a critical edition of Maestlin's Latin treatise on the nova of 1604 has been published (WLB, Cod. math. 4°, 15b. 11: *Consideratio Astronomica inusitatae Novae et prodigiosae Stellae, superiori 1604 anno, sub initium Octobris,*

---

1 See, for example, Ruffner (1971) and Heidarzadeh (2008).

2 See Maestlin (1573), (1578) and (1581), as well as Hellman (1944), Westman (1972), (2011) and Granada (2007), (2013).

3 Brahe (1602, III, 58).

4 Jarrell (1971, 127–128). See also Betsch (2002).

5 Jarrell (1971, 127).

iuxta Eclipticam in signo Sagittarii vesperi exortae, et adhuc nunc eodem loco lumine corusco lucentis [*Astronomical consideration of the extraordinary and prodigious new star that appeared near the ecliptic in the sign of Sagittarius one evening in early October in the preceding year 1604, and continues to shine in the same place with a tremulous light*]).[6] Now, in this volume, we offer an edition of the German manuscript treatise on the comet of 1618 (WLB Cod. Math. 4° 15b, Nr. 8) with an English translation. This Preface briefly describes the peculiar character and structure of the manuscript in the form in which it has arrived to us, together with a justification of the choices regarding the constitution of the text as it appears in the present edition.

In the Introduction, we describe the following issues: (Chapter 1) the broad impact of the three comets of 1618 on European astronomy on the eve of the Thirty Years' War (1618–1648); (Chapter 2) the genesis and structure of Maestlin's treatise on the third and greatest comet; (Chapter 3) the dates of appearance of the three comets and Maestlin's observations of them; (Chapters 4 and 5) Maestlin's earlier cometary treatises and the differences in their outlook with respect to the German treatise of 1618–1619; (Chapter 6) the Aristotelian Bartholomaeus Keckermann (ca. 1572–1609) and his fierce rejection of the interpretation of novas and comets as natural phenomena in the heavens; (Chapter 7) Maestlin's keen reply to Keckermann, and finally (Chapter 8) the introduction of sunspots and the telescopic observations of the Moon into the discussion of the celestial location of the comet and the ensuing elimination of cosmological dualism. A final chapter focuses on the issue of the predictability of comets and the claim of the Ulm mathematician Johann Faulhaber (1580–1635) to have predicted the appearance of the first comet of 1618 for 1 September, a possibility that Maestlin decidedly rejects.

The first part of this volume presents the transcription of the manuscript. It has arrived to us in two very different versions: a clean copy (Cod. math. 4° 15b, Nr. 8c), extending from pages 9 to 65, and a rough copy or first draft of the treatise (Cod. math. 4° 15b, Nr. 8a), spanning 27 pages and offering in the first eight an initial draft of the first sheet with the eight pages absent from the clean copy. The title of the treatise appears on the first page. The two copies of the manuscript treatise are accompanied by a set of five slips of paper (Cod. math. 4° 15b, Nr. 8b) presenting a first draft of the significant discussion with Keckermann in Chapter 8, still absent from the rough copy, as well as a first draft of several issues in the preceding chapters. These five slips of paper, writ-

6 See Granada (2014). The Latin text (106–116) is followed by an English translation by Patrick J. Boner (116–122).

ten on both sides and numbered I, II, 1–8, thus represent an intermediary stage between the two copies. The same slips, as well as pages 11–26 of the rough copy, allow us to follow Maestlin's reasoning from his first thoughts to the final formulation found in the clean copy. For this reason, reference is frequently made in the notes to both the slips and the second section (pp. 11–27) of the rough copy. Further particulars of the manuscript are presented in Chapter 2. In the second part, we present an English translation.

A first appendix in both parts offers a section, clearly very significant, missing from the clean copy but present in the rough copy, which discusses the question whether comets can be predicted (*An Cometae prognosci possent?*). This is precisely the issue that Faulhaber had raised, and about which Maestlin had been previously consulted by the Reutlingen mathematician Matthäus Beger (1588–1661), who acted as an intermediary.[7] The publication as an appendix reflects our conviction that, even though this section appears on pages 9–10 of the rough copy (Nr. 8a), it was probably too long to fit into the first missing pages of the clean copy (Nr. 8c). Accordingly, Maestlin most probably decided to approach the question in a later chapter of the last version, devoted to the meaning and effects of the comet. This chapter, however, was not written, since the only extant (autograph) copy breaks off abruptly before Chapter 9 is finished. Finally, a second appendix transcribes the final part—after 13 lines opening a short discussion on the meaning of the comet ("Was aber diser Comet bedeutte")—of page 27 in the rough copy, where there seems to be a draft of a letter addressed to the Duke of Württemberg, Johann Friedrich (1582–1628), justifying Maestlin's failure to deliver the report on the comet requested by the Duke. The date at the foot of the page, 9 February 1619, offers an approximate date for the conclusion of the incomplete draft, while the fact that it is the response to a request by the Duke ("bevelch", that is, "Befehl", order or command) explains why Maestlin writes in vernacular German, thus breaking with his normal custom of composing his mathematical and astronomical works in Latin.

The authors have distributed their work each according to his own abilities. Miguel Á. Granada, the primary author and principal investigator, has written the introduction and transcribed the manuscript, while Patrick J. Boner has translated the manuscript into English and aided with the preparation of the volume in various other ways. The notes to the English translation are the

7 Beger, a modest craftsman, was in contact from ca. 1607 with Maestlin, who lent him many books on mathematics. See Hawlitschek (1995, 31, 216–222), (2006, 109–113).

product of our joint endeavour. It goes without saying that both authors have contributed to the better performance of each other's work.

This edition is the result of research conducted by the project "Cosmología, teología y antropología en la primera fase de la Revolución Científica (1543–1633)," funded by the Spanish Government (Ministerio de Economía y Competitividad, Projects FFI2012–31573 for the triennium 2013–2015 and FFI2015–64498-P for the triennium 2016–2018). Miguel Á. Granada, who initiated research on Maestlin's manuscript, warmly thanks the Württembergische Landesbibliothek in Stuttgart for kindly allowing him to reproduce some pages from Maestlin's treatise on the comet of 1618 (Cod. Math. 4° 15 b, Nr. 8), as well as Wilhelm Schickard's beautiful figure of the comet's path in his manuscript treatise *Cometen Beschreibung* (Cod. math. 4° 43, fol. 157). His thanks also go to the library staff, especially Dr Kerstin Losert for her kind assistance and attention during the time he spent at the library. Miguel Ángel is especially grateful to Friedrich Seck, who carefully oversaw the final phase of the transcription of the manuscript. His knowledge allowed us to correct many erroneous readings as well as to clarify many doubtful ones, particularly in the transcription of the rough copy. Carlos Gilly and Jürgen Hamel have also been helpful in finally solving some obscure points in the transcription. Denis Savoie aided us in commenting on Maestlin's demonstration of the absence of parallax in the comet.

The work greatly benefited from research conducted by Miguel Ángel at the Center for the History of Science, Radboud University, Nijmegen in June 2015, following an invitation by Hans Thijssen and Christoph Lüthy. Miguel Ángel also thanks the Max–Planck–Institut für Wissenschaftsgeschichte in Berlin and its director, Professor Jürgen Renn, for warmly accepting and hosting him as a researcher during a sabbatical leave in 2014. To all of them, as well as Pietro Daniel Omodeo, Anna Jerratsch and other members of the Institut, we express our most sincere gratitude for their collaboration and suggestions.

Finally, we wish to thank Édouard Mehl and Friedrich Seck for their insightful reviews of the text presented to the publisher. Their final suggestions and corrections have allowed us to rectify many errors, improve the transcription of the manuscript at several points, and enrich the annotation with some significant additions. It is our hope that this edition will contribute to a better and more complete understanding of the important work of Maestlin, a titan in his own right who remains best remembered for introducing Kepler to the astronomy and cosmology of Copernicus. Needless to say, any remaining insufficiencies are entirely our own responsibility.

Miguel Á. Granada
Patrick J. Boner

Figures

Symbols of Planets and Zodiacal Signs

Sun	☉
Moon	☽
Mercury	☿
Venus	♀
Mars	♂
Jupiter	♃
Saturn	♄
Aries	♈
Taurus	♉
Gemini	♊
Cancer	♋
Leo	♌
Virgo	♍
Libra	♎
Scorpio	♏
Sagittarius	♐
Capricorn	♑
Aquarius	♒
Pisces	♓

Criteria of This Edition

- Double letters (ae, oe) and common abbreviations like those for -en and -er at the end of a word have been resolved.
- Underlining (always red in the handwriting) is reproduced in bold.
- Latin words appear always in italics.
- ⟨...⟩ indicates integration to the text.
- †...† indicates doubtful reading.

Introduction

Miguel Á. Granada

1 The Impact of the Comet of 1618 in Europe and Württemberg

1618 was an important year for cometary theory. It witnessed the appearance of three comets, the first ones that could be observed with the telescope, famously applied to astronomical observation by Galileo in 1610. As a consequence, and also because the previous comet had appeared in 1607, the number of technical or 'scientific' treatises written and printed in Europe reached new heights. If we recall that 1618 also saw the start of the Thirty Years' War, we may imagine even more vividly the abundant production of astrological, political and even eschatological tracts across the continent. Suffice only to mention the literature published in Bohemia, where the war first broke out, between 1618 and 1620, as studied by Vladimir Urbánek.[1] Germany had already been flooded with the same kind of cometary literature over the course of the sixteenth century.[2] The publication of the Rosicrucian manifestos in 1614 and 1615, along with the enthusiastic responses that dealt in detail with the celestial novelties (novas of 1572, 1600 and 1604) as harbingers of the imminent transformation of Europe, represented a significant contribution to this literature.[3]

The production of 'savant' literature soared all over Europe. In Italy, it coincided with the definitive adoption of the Tychonic world-system by the Jesuit order, in connection with Tycho's cometary theory,[4] and the protracted discussion between the Jesuit Orazio Grassi, a mathematician at the Collegio Romano, and Galileo that culminated in the latter's *Saggiatore* (1623).[5] In the Netherlands, discussion concerning the comets of 1618, brilliantly studied by Tabitta Van Nouhuys,[6] produced the important treatise of Willebrord Snel, whose printed edition in 1619 included Christoph Rothmann's treatise on the

1 Urbánek (2002).
2 Barnes (2016). See Jerratsch (2020) for the *Flugschriften* printed in German.
3 Barnes (1988) and Gilly (2012).
4 Besomi and Camerota (2000), Baldini (1992, 183–281). See also Lerner (1995).
5 Drake and O'Malley (1960). See also Galileo–Grassi (2016). Independently of the literature produced by the Italian Jesuits and the Galileo–Grassi controversy, Giovanni Camillo Gloriosi, successor to Galileo in the chair of mathematics at the University of Padua, also published a long and significant treatise (Gloriosi 1624) aligned with the positions of the 'more recent astronomers'.
6 Van Nouhuys (1998).

 | DOI:10.1163/9789004512641_002

comet of 1585.[7] Following on from the contemporary debate in Germany, the same comet attracted a great deal of attention in France, where it led to the later emergence of the Cartesian 'fable du monde', as Édouard Mehl has persuasively argued.[8] In Portugal in 1619, Manuel Bocarro Francês (alias Jacob Rosales, when he re-embraced Judaism in exile) published a *Tratado dos cometas que apareceram em Novembre passado de 1619*, which defended a celestial location for the comets in the framework of Tychonic cosmology.[9] Bocarro met strong opposition from conservative Aristotelians, however, and the traditional interpretation of the comet as a sublunary phenomenon, along with an emphasis on astrological aspects, characterised the literature published in Spain as well.[10]

Finally in 1619, Germany witnessed the publication of a vast work by Johannes Kepler (1571–1630), *De cometis libelli tres*,[11] as well as a myriad of tracts of every sort, recently studied in part by Marion Gindhart and Anna Jerratsch.[12] Among them, the treatise by the Jesuit Johann Baptist Cysat, disciple and assistant of Christoph Scheiner at the University of Ingolstadt, deserves mention. His *Mathemata Astronomica de loco, motu, magnitudine et causis cometae qui sub finem Anni 1618. et initium Anni 1619. in coelo fulsit*, published in 1619, reveals the transfer in the Jesuit Order, already underway, from Aristotelianism to Tychonic cosmology as well as the application of the telescope to the observation of the head of the comet.[13] Many other treatises of very different kinds and of varying importance have remained in manuscript all over Europe and still await study.

One of the most prolific German territories, in terms of both printed and handwritten cometary studies, was Protestant Württemberg. There, in addition

7 Snel (1619). For Rothmann's treatise, see now Granada, Mosley, and Jardine (2014).

8 Mehl (2013) (2019). The comet was also observed in France by Pierre Gassendi between 29 November and 13 January 1619. His observations and conclusions regarding a measure of parallax of the comet smaller than that of the Sun were published in 1630. As a consequence, Gassendi located the comet above the orb of the Sun. See Gassendi (1630), pp. 319–328.

9 Bocarro (2009) and Carolino (2012).

10 See Rosselló Botey (2000, 42–47), Navarro Brotóns (2012), 38–40, Cedillo Díaz (2019, Appendix 2).

11 Kepler (1619).

12 Gindhart (2006), Jerratsch (2020, 323–383).

13 Cysat (1619, 72–75). See also Siebert (2006, 316–325). Johannes Remus Quietanus (1588–1654), who was in contact with Kepler, Galileo and the Jesuits in Rome and Ingolstadt, wrote a short treatise on the comets of 1618; see Remus Quietanus (1619). We have not been able to access a copy of this edition, which does not appear to be housed at VD17. A manuscript copy is included in the Galilean manuscripts of the Accademia dei Lincei and available at https://bibdig.museogalileo.it/Teca/Viewer?an=953307 (pp. 32^{r}–38^{v}). On Remus's relationship with Kepler and Galileo, see Bucciantini (2003, 207–209, 242–244, 275–278).

to the 'Kometenstreit' in the free imperial city of Ulm, sparked by the alleged prediction of the appearance of the first comet by the mathematician Johannes Faulhaber (1580–1635), along with the enthusiastic political and eschatological overtones of the Rosicrucian manifestos, two handwritten reports and interventions by two prominent mathematicians from the University of Tübingen were produced, both in German. The first one was written by Michael Maestlin (1550–1631), who became Professor of Mathematics at the University of Tübingen in 1583. Maestlin's manuscript treatise, titled *Astronomischer Discurs von dem Cometen, so in Anno 1618, im* Nouembri *zu erscheinen angefangen und bis im* Februario *dis 1619 Jars am Himmel noch gesehen wirt*, is the subject of the present volume. Preserved at the Württembergische Landesbibliothek in Stuttgart (WLB, Cod. Math. 4° 15b, Nr. 8), this treatise, written when Maestlin was already 68 years old, was left unfinished, extending over 65 pages, and as a consequence was never delivered to the petitioner, who, as we shall show, was none other than Duke Johann Friedrich of Württemberg (1582–1628). Maestlin's manuscript, whose first pages are written in an almost illegible hand, has not been studied to date, even though its existence was known to Maestlin scholars.[14] Recently, after a previous edition and study of Maestlin's Latin manuscript on the nova of 1604 in Serpentarius,[15] we published an initial study of his treatise on the comet of 1618 as a first step in a broader project that culminates with the present edition and translation.[16]

The second manuscript, written by the young Wilhelm Schickard (1592–1635), was a long treatise, offering a complete cometary theory and extending to an examination of earlier comets. It was titled *Cometen Beschreibung In zwen underschidliche Partes abgetheilt, deren Erster Von denselbigen ins gemein: der Ander Von allen Insonderheit, sonderlich aber denen drey Jüngsten, Im abgeloffnen 1618. Jahr erschienen, aussführlich handelt.* A sumptuous copy, splendidly illustrated with aquarels in vibrant colours by Schickard himself, is also preserved at the Württembergische Landesbibliothek and has been digitally reproduced and made accessible online.[17] When the comets of 1618 appeared, Schickard was a Lutheran minister in Nürtingen, a small town near Tübingen. His magnificent treatise—also commissioned by Duke Johann Friedrich of

14 Jarrell (1971, 127–128), Betsch (2002). As Jarrell argued on p. 127, the alleged publication by Maestlin of a treatise on the 1618 comet, still accepted by Hellman (1944, 142), is a "bibliographical ghost". For a short, but very accurate presentation of Maestlin's work in the Tübingen context, see also Jarrell (1981).

15 Granada (2014).

16 Granada (2016). We rely on this essay extensively in the present Introduction.

17 WLB, Cod. math. 4° 43 (http://digital.wlb-stuttgart.de/purl/bsz307044173). We intend to publish the transcription of Schickard's treatise in a future volume as well.

Württemberg on 18 February 1619 after Maestlin failed to comply with his own report—and the positive impression it made on the Duke could have contributed to his nomination in May 1619 as Professor of Hebrew at the University of Tübingen.[18] Schickard's presentation copy of his treatise to the Duke, covering 174 pages written in a beautiful hand with clear spelling and adorned, as indicated above, with a great number of images, has attracted some attention,[19] but by no means the full consideration it doubtless deserves.

2 Maestlin's Treatise on the Comet of 1618

Maestlin devoted a great deal of attention to the celestial novelties of his time. He wrote and published two significant treatises, one on the nova of 1572 in Cassiopeia and another on the comet of 1577. These two treatises are among the most significant and innovative contributions of the period and won the admiration and respectful commentary of Tycho Brahe (though the Dane also criticised Maestlin's Copernican views).[20] Maestlin published a third treatise on the comet of 1580, but the quality of this study was considerably inferior and, as a consequence, has attracted far less attention from historians. Although he published nothing on any later comets or novas, Maestlin always paid considerable attention to them[21] and even wrote, as indicated above, two incomplete

18 Schickard's presentation copy is dated 15 April 1619 (see note 29 below). His nomination was under negotiation from November 1618. After Maestlin's death in 1631, Schickard was appointed to the chair of astronomy. His career closely mirrors that of Maestlin, who was also a deacon for some years and became Professor of Mathematics at the University of Heidelberg only in 1580, after publishing his significant treatise on the comet of 1577. We may compare the careers of Maestlin and Schickard to those of other scholars whose cometary studies may have earned them an academic appointment. Five weeks after first witnessing the comet of 1577, Danish astrologer Jørgen Dybvad published a pamphlet on it that, in combination with support from the Duke of Saxony, secured him the chair of mathematics at the University of Copenhagen. On the work of Dybvad and his role as "a dangerous rival" to Brahe, see Christianson (1979).

19 See Mauder (1995) and Schramm (1978, 175–178, 273–274).

20 For Maestlin's work on the nova of 1572 and Brahe's criticism, see Granada (2007, 2013). For his work on the comet of 1577, see Westman (1972).

21 An enthusiast of comets and a friend of Maestlin, Helisaeus Roeslin was notified by Maestlin of the appearance of the comet of 1585, which was scarcely visible. Roeslin had not yet detected the comet; on this important episode, see Roeslin (1597, 16^v) quoted in Granada (2002, 294). For his own part, Isaac Habrecht (1589–1633), a physician and mathematician from Strasbourg, also mentions some opinions of Maestlin concerning the comets of 1596 and 1607; cf. Habrecht (1618, 47).

treatises: the first on the nova of 1604,[22] and the second precisely on the comet of 1618.

A colleague of Maestlin and jurist and philologist at the University of Tübingen, Johann Bernhard Unfried (1589–1635), recalled in a personal manuscript codex on various subjects[23] that Maestlin had been asked by another colleague, Heinrich Welling (1555–1620; professor of classical philology at Tübingen), about the comet that appeared in November: "Question: When did the comet appear in this year 1618, in what constellation and in what degree? What is its origin and its motion?" Unfried then recorded Maestlin's response, dated 1 December 1618, the earliest known statement by the Tübingen astronomer about the new phenomenon, which lends this (hitherto neglected) testimony special interest:

> Answer: This comet travels through Arcturus. I don't know when it appeared. Several days ago, I saw a tail at the time of morning, though it did not seem to me to be anything like a comet; rather, I reckoned it was a ray of twilight or dawn of the new day, especially since its head did not then appear, but was hidden by clouds on the horizon. Its motion is nearly straight from the south towards the North Pole. However, I have not seen it now for several mornings, or even if I have glimpsed it through the clouds, I have not been able to observe it astronomically. M[aster] Michael Maestlin, Mathematician at Tübingen, 1 December 1618. To Heinrich Welling, distinguished Vice-Dean and his colleague.[24]

---

22 For an edition and English translation of this manuscript treatise, see Granada (2014).

23 J.B. Unfried, *Kollektaneen*, Landesarchiv Baden-Württemberg, Abteilung Hauptstaatsarchiv Stuttgart, J1 Nr. 23, p. 1497.

24 This is the German text of this important testimony, whose data coincide with Maestlin's observations of the third comet of 1618: "Comet im November anno/1618/ Quaestio./ Wan der Comet, in qua Constellatione, und in quo gradu, in anno 1618, erschinen?/Und was sein principium und motus seye? // Responsio./ Cometa hic versatur circa Arcturum: quando apparere coeperit, nescio. Ich hab zwar vor etlich tagen zuvor caudam gesehen tempore matutino, hab aber mich nichts wenigers, als eines Cometes versehen, sonder vermeint, es were radius crepusculi seu diluculi ex orituri diei, sonderlich die weil sein caput dazumal nicht erscheinete, sed nubibus ad Horizontem occultatum fuit. Motus eius est ab Austro propemodum directè versus polum Mundi. Doch hab ich ihne nun etlich Morgens nicht gesehen, oder ob ich schon per nubes ihn gespürt, hab ich ihn doch Astronomicè nicht observieren könden. M. Michaël Maestlinus, Mathematicus Tubingensis, Calendis Decembribus anni 1618. /ad Henricum Wellingium, spectabilem pro-Decanum et Collegam suum".—Below is noted a "Prague Notice of December 3./ The comet, which appears here at 5 p.m. and is visible until one hour before daylight, should have a longitude of 9 degrees and each degree has 15 cubits./ To this Master Maestlin added in the margin:/

It was no secret that Maestlin was writing on the comet. On 1/11 January 1619 the Tübingen jurist Christoph Besold (1577–1638) wrote to Kepler with the following request: "I wish to see your opinion on the comet, for it will doubtless be published through the press. Our Maestlin observes his old habit: he doubts, fears and delays, and has still published nothing, though he labours on a large and complete treatise on comets".[25] Some days later, Maestlin himself affirmed in a letter to Faulhaber that he intended to publish on the comet shortly: "[…] such as, God willing, shall soon be specifically shown".[26]

The resulting treatise, however, may come as a great surprise to the reader because it is written in German. Maestlin's scientific production, and particularly his astronomical works and earlier treatises on novas and comets, were all written in Latin. There must have been a reason for this deviation, and the final page of the rough copy seems to provide some clues. The treatises of 1578 and 1581 were dedicated to Duke Ludwig of Württemberg (1554–1593) on Maestlin's own initiative, without any previous request for a report, as it seems, from the prince.

However, page 27 of the first rough section[27] contains a telling passage (dated 9 February 1619 at the bottom), which is unfortunately very difficult to read. The passage looks like a first draft or some notes written for a letter regarding the still incomplete treatise. It is easy to make out the repeated formula 'e. f. g.' (*eure fürstliche Gnaden*, that is, "your princely Grace"), doubtless referring to the Duke, as well as the word 'Bevelch' (*Befehl*, 'order' or 'command'), indicating that the request for a written report very probably stemmed from Duke Johann Friedrich. This report had to be written in German, as was customary,[28] and as Wilhelm Schickard also did with his *Cometen Beschreibung*.[29]

This mathematical observer was Lippus. Sunday, December 6, 1618". We thank Carlos Gilly for informing us of this important document.

25 *JKGW*, xvii, nº 822, 313: "Aveo porrò videre tuum de Cometa judicium, sine dubio enim praelo vulgatum erit. Maestlinus noster antiquum obtinet; dubitat, timet et cunctatur, nec aliquid edidit: quamvis magnum et integrum de Cometis tractatum moliatur". This letter has already been indicated by Jarrell (1971, 127).

26 Letter of January 1619: "[…] wie, ob Gott will, soll in kürtze, specificè angezaigt werden". This significant letter, which shall be referred to later, has been recently edited by Mehl (2013, 254–255).

27 See the transcription below, p. 194, Appendix 2 to Maestlin's treatise.

28 Thus, for example, Philip Apianus (1531–1589), Professor of Mathematics at the University of Tübingen when the nova of 1572 appeared, wrote his report for Duke Ludwig in German (dated 26 December 1572). This report is preserved in the Württembergisches Hauptstaatsarchiv Stuttgart, Bestand A 274 Bü 21.

29 See the dedicatory letter, dated 15 April 1619, in Schickard (1619, 4^r–5^v): "[…] durch Ihre F[ürstliche] G[nade] mir gnedig anbefehlen lassen, von solchen Cometen mein fernere

Nevertheless, and apparently in contrast to Schickard, Maestlin did not comply with the request, thus justifying Besold's concerns. Maestlin's treatise was quite long but never finished, as had happened with his previous piece on the nova of 1604. Most probably, the reason was that from 1602 Maestlin had suffered from periodic bouts of depression and was described by contemporaries as *melancholicus*.[30] He frequently withheld excessive time or failed to return manuscripts that had been submitted to him for review or evaluation, and after 1610 he published nothing apart from a reprint of his *Epitome astronomiae*.

The autograph manuscript has come down to us in two very different sections: the second part or section (fascicle 8c) is written in a very clear and polished fashion, but it begins on page 9 and finishes on page 65, apparently without ever reaching a conclusion. Since the usual calling at the bottom to the beginning of the first word in the following page is missing, it is less likely that the following pages are lost. I am inclined rather to think that Maestlin did not continue writing and left it unfinished, just as he had done with his earlier Latin treatise on the new star of 1604. The reasons for his incompletion are unknown, but, besides Maestlin's old age and chronic *melancholia*, Besold's letter quoted above may give us an indication.

On page 11 in this clean copy begins the third chapter (*On the course of this comet, and the signs and constellations through which it passed* [*Von dis Cometen Lauff, welche Zaichen, und himlische Bilder Er durchwandert seye*]). We can thus infer that the two preceding pages belong to the second chapter. The first eight pages, and with them the title of the treatise, are lost.

By contrast, the first part or section of the manuscript (fascicle 8a), extending over 27 pages, comprises a rough copy. It is clearly a draft that is very difficult to read, including many deletions and parallel texts (which are even more difficult to read than the texts they replace). It provides us, at least, with the title of the treatise on page 1: *Astronomischer Discurs von dem Cometen, so in Anno 1618, im* Nouembri *zu erscheinen angefangen und bis im* Februario *dis 1619 Jars am Himmel noch gesehen wirt* [*Astronomical Discourse on the Comet that First Appeared in November 1618 and Can Still Be Seen in the Sky in February of this Year 1619*]. See Figure 1.

Meinung einfeltig schrifftlich zu verfassen, und E[ure] F[ürstliche] G[nade] underthönig zu ubergeben", p. 4^{v}. See also Mauder (1995, 153).

30 Thus, in a letter of 12 January 1603 to Herwart von Hohenburg (*JKGW*, xiv, nº 242, 473–475), Kepler reported that Maestlin had descended to melancholy when his son had sought refuge with the Jesuits. "The truth is," Kepler wrote, "Maestlin was stricken by melancholy over the flight of his son, whom they say lies in hiding among the Jesuits." On this, see Seck (2002, 118–121).

However, these preliminary 27 pages do not in their entirety constitute a preliminary version of the first set of eight lost pages in the clean copy. In fact, most pages in this first, rough section present a text similar to other pages found in the final copy (fascicle 8c). Nonetheless, the preliminary eight pages of this section seem to offer a first version of the missing eight pages in the final copy. Thus, after the first introductory page, we find on page 2 the beginning of the first chapter with the title (Caput I. *In verschinem 1618 seind mehr als nur ain Comet erschinen* [*More than one comet appeared in the previous year 1618*]). Similarly, on page 7 (a page in which the text is very convoluted) Chapter 2 begins, with the title indicated in the margin (II. Cap. *Von dises Cometen* [the third] *ersten Aufgang und Erscheinung* [*On the first emergence and appearance of this comet*]); see Figure 2. From all this, we may reckon that in the following rough pages we find the transition to pages 9 and 10 closing the second chapter in the clean copy.

Interestingly, on page 9 of the first version an addition in the margin asks, "Can comets be predicted?" (*An Cometae prognosci possent?*), indicating the content of the page, whose subject continues on the following page. Maestlin gives no name to indicate the possible target of his resolutely negative response, and it may be asked whether the issue was addressed in the second chapter of the final version. It is most probable that Maestlin finally decided not to discuss this question at this point and preferred to leave it for a later section, possibly a final chapter on the significance and meaning of the present comet. For this reason it can be persuasively argued that the transition to the clean copy is found on page 8 in the rough copy. Moreover, page 11 of the rough copy presents in the upper left margin the title "Motus via", indicating a development that coincides until the end of page 12 with the text of Chapter 3 in the final version. A comparison of the remaining pages in the rough copy with the rest of the final version shows that these preliminary pages are a first draft of issues addressed in the clean copy, with the exception of pages 39–58 in Chapter 8 (containing the important discussion with Bartholomaeus Keckermann and the digression on sunspots), which are not anticipated in the rough copy. Accordingly, it may be argued that Maestlin decided at a later time to introduce a long criticism of Keckermann's attack on the celestial character of comets. This is confirmed by the presence in the manuscript, between page 27 of the rough copy and the beginning of the clean copy, of a set of five pages (fascicle 8b, numbered by a later hand I, II, 1–8), of a shorter format and written on both sides. These pages, more precisely described as slips of paper, contain a first draft of issues still absent from the rough copy, among them (on pp. 1–2, 5 and 7–8) a first draft of the discussion with Keckermann and of the conclusions to be drawn from the recent telescopic observations. In the notes to the translation (in the second

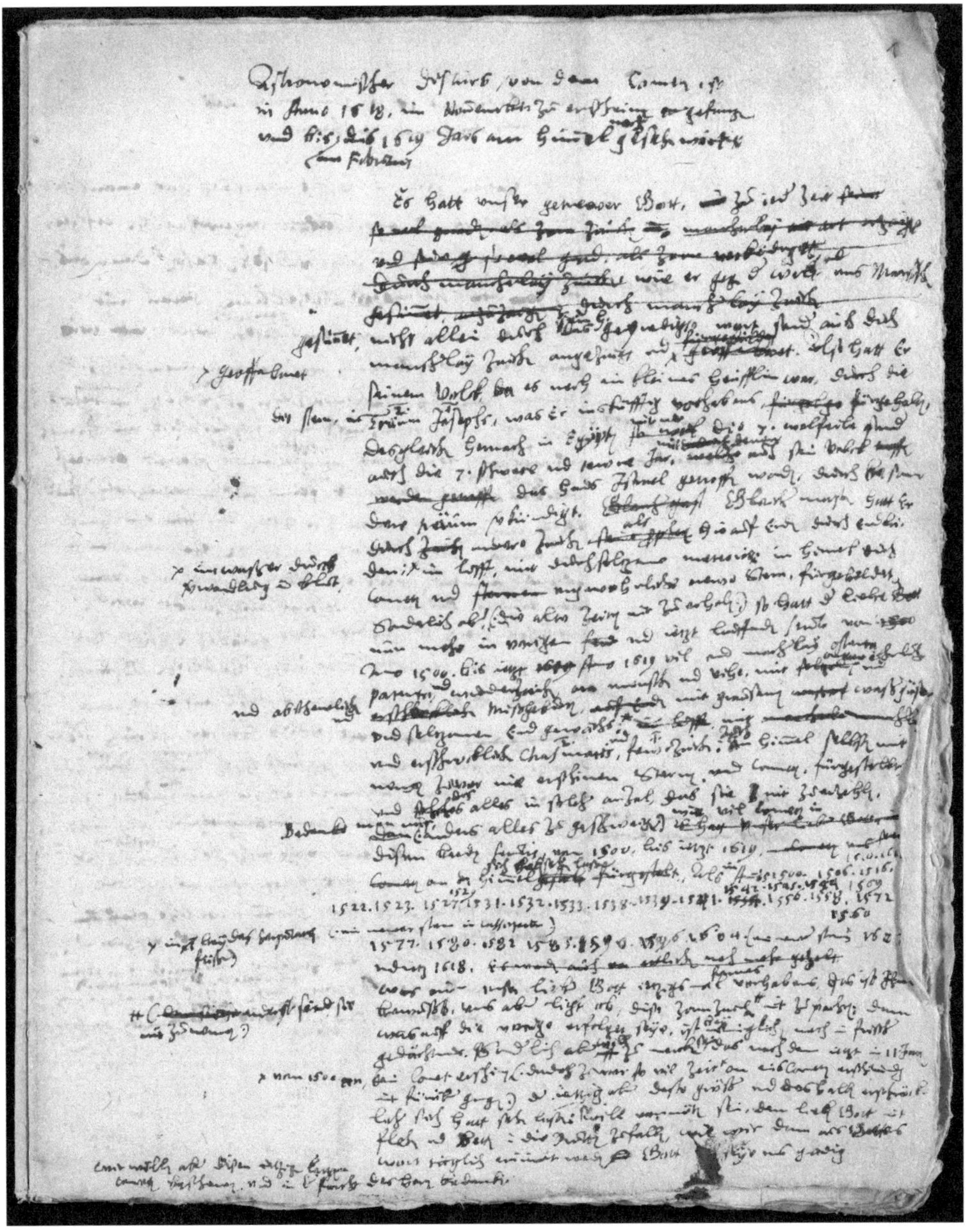

FIGURE 1 First page of the fascicle 8a showing the title of the treatise

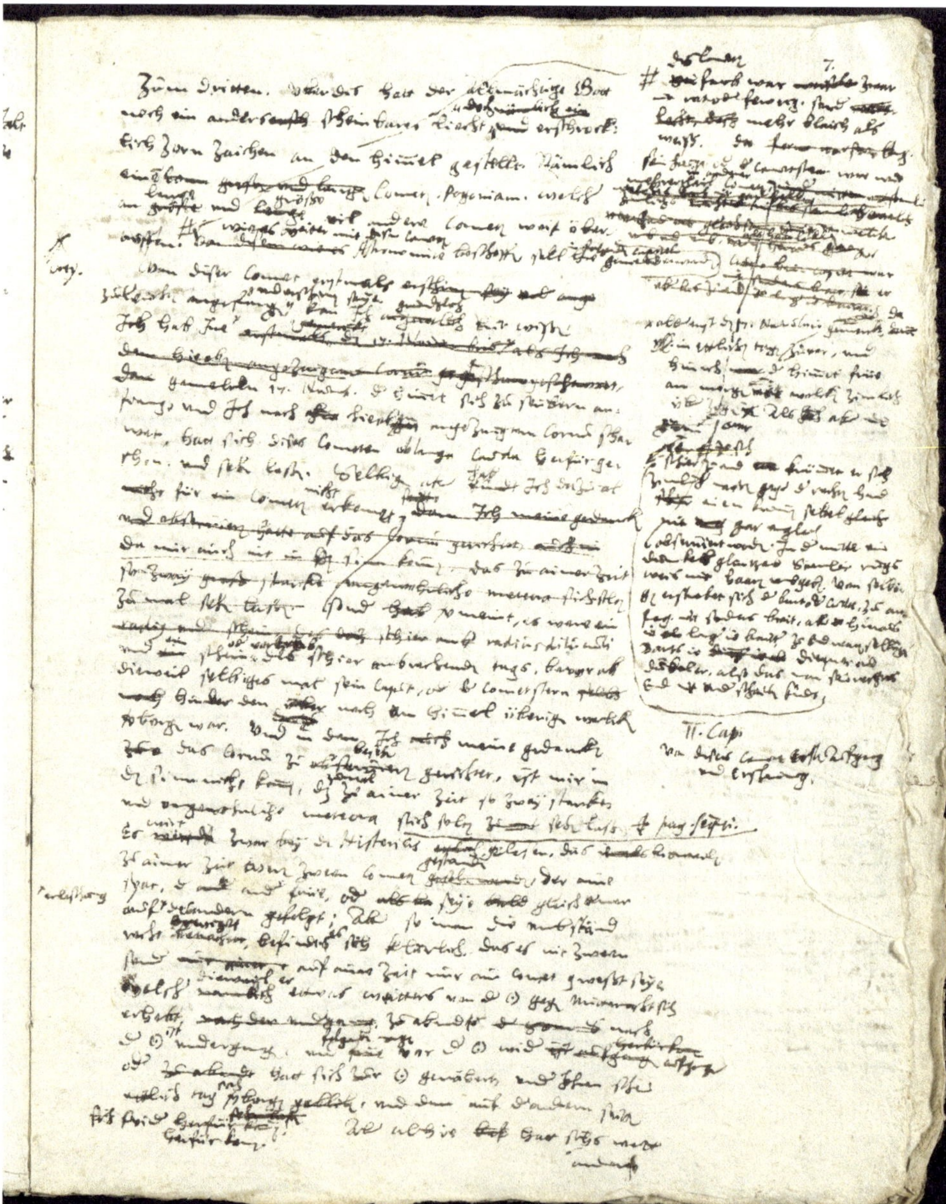

FIGURE 2 Page 7 of the fascicle 8a showing the beginning of the second chapter

part of this volume), we frequently indicate the connections, both in the rough copy and in this fascicle 8b, with passages in the clean copy, as well as the occasional presence of some significant passages that were ultimately omitted in the final writing.

As a consequence, the edition of the rough copy is limited to the first 10 pages that allow for a reconstruction of the missing first eight pages of the clean copy, in addition to the question (in Appendix 1) of the predictability of comets.

Unfortunately, Maestlin did not complete the final version up to the chapter on the significance and predictability of comets, since the manuscript breaks off before he addresses it.[31] The issue is very significant, however, for it may refer to the claim that Johannes Faulhaber had predicted the appearance of the comet in 1617 by virtue of his mystical numbers. This is all the more probable, as there was an exchange between Faulhaber and Maestlin on the subject. The Tübingen professor rejected such a possibility in his letter to the Ulm mathematician, referring Faulhaber to a future intervention, perhaps (as suggested) an allusion to his treatise on the comet. This issue is discussed below, in Chapter 9 of this Introduction. For the moment, I merely indicate the titles and pages of the following chapters of the treatise in the final version (fascicle 8c). They are as follows:

> 3. On the course of this comet, and the signs and constellations through which it passed [Von dis Cometen Lauff, welche Zaichen, und himlische Bilder Er durchwandert seye], pp. 11–15.
>
> 4. That the philosophers are divided in their opinion over whether comets are elementary or ethereal, that is, whether they are generated and brought into being here below in the air or high above in the heavens [Das die *Philosophi* wideriger Meinung sind, ob die Cometen *Elementares* oder *Aetherei* seien, das ist, ob sie hie unden im Lufft, oder droben im hohen himmel generiert und geschaffen werden] pp. 15–17.
>
> 5. Whether and how we may find a solution for the two opposing opinions [Ob und wie in den wider wertigen Meinungen ein Außschlag zu finden seye], pp. 17–22.

31 It can safely be said that, even if Maestlin's treatise had contained a chapter devoted to an interpretation of the meaning of the comet, that is, a consideration of its effects, this chapter would not have gone very far, given Maestlin's aversion to astrology and his refusal to make astrological predictions, already explicit in his treatises on the comets of 1577 and 1580. For his own part, though cautiously, Schickard devoted a significant portion of his treatise to prognostication; see Schickard (1619, 162–174) ("Von seiner Bedeuttung").

6. Whether our present comet possessed any sensible parallax or not, and how far away it may have been from the Earth [Ob diser unser Comet einiche *parallaxin sensibilem* ghabt habe, oder nit: und dann wie weit er von der Erden gestanden möchte sein], pp. 22–26.

7. That before this time many other comets appeared and were observed not in the air, but in the upper heaven [Das auch vor diser zeit vil andere Cometen nicht im Lufft, sonder im hohen himmel erschinen, und observiert worden seind], pp. 26–35.

8. What Aristotle and other philosophers might have been missing that led them to think about comets the wrong way [Wa es dem *Aristoteli* und andern *Philosophis*, von den Cometen anderst zu halten, als es sich befindt, gemangelt habe], pp. 35–53.

9. Several questions concerning comets in general, and what follows from them [Ettliche fragen, die Cometen in gemein, und was inen anhengig, betreffend], pp. 54–65].

3 The Comets of 1618–1619 and Maestlin's Observations

1618 saw the appearance of three comets: the first, visible in August and September, bears the astronomical identification C/1618 Q1; the second, visible in November and the beginning of December, is known as C/1618 V1 and also designated as 1618 III. Both were barely visible for a few weeks and attracted the attention of only a small number of observers. The third and final comet, by far the superior of 1618 and the main subject of Maestlin's treatise, was a very great comet, visible from November to January (according to Maestlin, also in February and even as late as March) and is known as 1618 II and C/1618 W1.[32]

In the most important treatise published in Germany, Kepler described the appearance of the three comets of 1618 in the following terms: "the first, which arose in the month of August, was most dark throughout September, so that it was scarcely noted even by the most attentive astronomer, not to mention by common people. The second and third were observed in November: the second [C/1618 V1] only by its tail and not beyond the end of November; the third [C/1618 W1], both by its head and its tail, most clearly through the whole of December and, like other [comets], until a good part of January".[33] According

32 Kronk (1999, I, 333–341).

33 Kepler (1619, 177; our translation). For the full observational history of the three comets as described by Kepler, see *ibidem*, 177–198. For Schickard's observations, see Schickard (1619) 143–146 (first comet), 146–148 (second comet) and 149–158 (third, great comet).

to Gary Kronk, the most probable dates for their appearance and disappearance were 25 August–25 September for the first comet, 11 November–9 December for the second and 25 November 1618–22 January 1619 (dates according to Universal Time) for the third.[34] Consequently, the second and third comets were seen simultaneously in the sky, not very far from one another, thus giving Kepler occasion to hypothesise that it was one and the same comet broken in two.[35]

Maestlin addresses the above issue in the first chapter of his treatise, preserved only in the rough copy. With regard to the first comet, Maestlin confesses that neither he nor anyone else in Tübingen was able to see it:

> First, in August 1618, as I have been informed, a comet should have appeared in the sky under Ursa Major [...]. I have nothing to present concerning this comet, since I did not catch sight of it. [...] And it comes as no little surprise that this comet was not observed around here even by commoners, particularly by the field guards.[36]

Maestlin observed the second comet [C/1618 V1] from 10 November (o.s.) as a brilliant *horn* (the tail) over the horizon and assumed that its head was visible only to people living in more meridional latitudes. He was not sure whether the

34 Kronk (1999, I, 333–341). See also Kapoor (2016, 267–268).

35 Kepler (1619, 178. 27–30): "Video duos hosce passim in unum confundi: nam et eodem tempore sunt visi, et ab eodem loco coeli progressi: ut valde opiner, de iis idem esse factum, quod de uno sui temporis prodidit Ephorus ad Senecam [*Naturales quaestiones*, VII, 16, 5], unum in duas partes dissiliisse, inque diversas mundi plagas abiisse". See also *ibidem*, 179. 38–180. 1; 184. 19–25. In his letter to Maestlin of 3 December, Kepler wrote that he had seen for the last time the second comet on 19/29 November, the same day he first saw the third: "Alterius caudam solam, nec eam totam vidimus a 10/20 in 19/29 Nov. [...] Tertius totus [...] effulsit 19/29 Novembris, ut *uno tempore duos cernerem*", *JKGW*, xvii, letter nº 815, lines 346–351; emphasis ours. This is in agreement with *De cometis libelli tres*, *JKGW*, viii, p. 185. 27–186. 3.

36 Maestlin (1619, p. 3 rough copy) "Und Erstlich in verschinem *Augusto* des 1618. Jars, solle, wie Ich berichtet werde, ein Comet sich am himel *sub Ursa maiore* [...] habe sehen laßen. Von disem Cometen hab Ich nichts fürzubringen, dann er ist mir nicht ins gesicht kommen. [...] Und nimpt mich nit wenig wunder, das bey uns alhie diser Comet nicht auch ist von gemeinen leutten, sonderlich von den feldschützen animadvertirt worden". Contrary to Maestlin, Schickard affirmed (see Schickard (1619, 144)) that he was made aware of the comet by some peasants and could observe it twice afterwards. Their different accounts suggest that there was no communication between the two. However, in a letter to Kepler of 27 December 1618 (*JKGW*, xvii, nº 820, lines 52–54) Schickard affirmed that he never saw this first comet: "Quod primum attinet, sub Ursa, equidem, ut ingenue fatear, nunquam illum vidi, licet famâ eius excitatus, aliquoties ante diem sedulò coelum contemplatus fuerim".

phenomenon could be rightly termed a comet and preferred to interpret it as a meteor of the kind called a *horn* or *cornu*. He observed it until 25 November (o.s.):

> Whether this bright light was also a comet, as many wish to call it, I freely leave to those who saw it, since it did not entirely appear altogether in our land, for its beginning, head, or comet star never emerged above our horizon [...]. In my opinion, this meteor should rather be called a horn, a horned star, or κερατοειδής, as the meteorologists distinguish such phenomena.[37]

In the first pages of the clean copy belonging to the second chapter, Maestlin notes that many observers identified it erroneously as the third comet and thus inclined to concede an earlier appearance to the final, third comet, unaware of the coincidence of the two phenomena in the sky:

> an error occurs among those who tell themselves they first became aware of this comet several days before (but in the morning). For what they saw could not have been the comet, but rather was the other meteor visible at the time, namely the horned star. Yet when the comet also arose and they did not notice that there were two *meteora* present in the sky at the same time, they thought the same horned star was also the comet.[38]

---

37 Maestlin (1619, 3–4 rough copy): "Ob diser glantz, auch ein Comet, wie etlich wöllen, zu nennen seye, setze Ich iedem, der so ine gesehen hatt, frey, sintenmal er nicht gantz und volkommen in unsern landen erschinen ist, sonder sein anfang, *caput* oder Comet stern ist über unsern Horizontem nit herauf kommen [...]. Meins erachtens solte dis *Meteor* vil mehr *Cornu*, κερατίας, oder κερατοειδής, wie die *Meteorologi* solche *phaenomena* underscheiden, genennet werden". See also p. 6: "As soon as I found out about it, I observed it every morning from 10 November to 25 November, as long as the sky was clear, and noted that it diminished daily". For a brief overview of the many ancient and medieval terms for comets according to their various colours and shapes, see Dall'Olmo (1980, 16–20).

38 Maestlin (1619, 11 in the clean copy). Maestlin, then, accepts the coincidence of the two phenomena, but, unlike Kepler, interprets the second comet as a meteor and does not contemplate the possibility that this comet was broken into two pieces. For his own part, Schickard did not believe that two comets could appear and be present simultaneously in the sky; cf. *Cometen Beschreibung*, section 28 ("Es seind noch nie zwen Cometen zu mahl am Himmel gestanden") in Schickard (1619, 81–84): "Ich rede iezt nit von underschidlichen Zeitten eines Jahrs, dan das ist gewiß, daß ettwan in kurtzer Zeitt zwen Cometen auff einander gefolgt sein. Doch wan der Eine vergangen, daß der ander sich erzeigt. Gestelt dann eben in disem 1618. Jahr, nit nur ein Comet gestanden, sonder drey underschidliche nacheinander sich erzeigt haben [...] Sonder Ich red iez von einerley Zeitt, daß nie zu mahl zwen gestanden, und beede zu mahl geleuchtet hetten, oder ein anderer kom-

The third phenomenon [C/1618 W1] was truly a comet, a "terrible sign of God's wrath": a comet with a long tail,[39] which Maestlin observed for the first time on 25 November (o.s.).[40] The last observation registered by Maestlin is that of 13 January,[41] five days before the comet disappeared, according to Kepler. Nevertheless, the title of Maestlin's treatise still located the comet in the heavens in February, and on page 13 of the clean copy even in March: "it finally spent several days from February until March not far from its left ear (*Ursa Maior*, star no. 6), as if stationary or standing still".[42]

Maestlin discusses his observations of the third comet in Chapter 3 (fascicle 8c, pp. 11–15). He clearly shows less concern over the matter and elaborates less on the observational results than in his earlier treatise on the comet of 1577. It is worthwhile examining this difference and considering the reasons for it, since they are connected with the differences in context, both astronomical and broadly philosophical, of the two treatises, separated by a span of forty years.

---

men wer, ehe dan der alt verschwunden", p. 81. Accordingly, Schickard initially interpreted them as the same comet (a common opinion, as reported by Maestlin, p. 11) and, following Kepler, finally came to see the third phenomenon as a new comet after the second had vanished. Cf. the brief description of the second comet in Schickard (1619, 146–148): "[I was] lang in der opinion gewest, es were kein newer, sonder der vorig, und eben diser andere Comet, seithero under den Wolkhen so weit fortgeruckt; sonderlich dieweil die Lini seines Lauffs zimlich wol mit einstimmete. Ist mir aber solche Meinung vom *Keplero* benommen worden, der zu Lintz damahls hell Wetter gehabt, und sein, des anderns Abnemmen und Undergang wissentlich wol gemerkt und mir zugeschriben hatt. Drumb hieran nit zu zweifeln, es sey der leste langwürige, ein sonderlicher eigner Comet, hab mit disem andern nichzit zu thuon", p. 148. No letter from Kepler to Schickard from this time is extant, but in the letter of 27 December to Kepler, Schickard seems to note Kepler's conception of a comet broken in two parts: "Porro illud miror, quod ex hoc uno, adhuc spectando Cometa binos facias" (*JKGW*, xvii, letter nº 820, lines 46–47).

39 Maestlin (1619, 7 in the rough copy): "Besides, Almighty God situated yet another apparent light, but truly a terrible sign of wrath in the sky. This was a comet, or bearded star, which far surpassed many other comets in length and size". According to Habrecht (1618, 149) the tail covered 40 degrees. Schickard registered the same length in his observation of 26 December; see *Cometen Beschreibung*, 149.

40 Maestlin (1619, clean copy, Chapter 3, 12): "den 25. Nouemb. war der Comet von mir observiert". Schickard also saw it on that date for the first time. Cf. *Cometen Beschreibung*, 149 and 149–162 for the minute description of his observation, accompanied by many beautiful images.

41 *Ibidem*: "Also den 13 *Ianuarii* fand ich in *per observationes in* 27. gr. 50. scr. ♋, von der *ecliptica* 57. gr. 27. scr.".

42 Fascicle 8c, p. 13. This also agrees with Maestlin's comments in his draft of the presentation letter on p. 27 in the rough copy (see Appendix 2 to Maestlin's treatise).

4 The Treatises of 1578 and 1580: A Mathematical and Astronomical Approach

We shall concentrate on the first cometary treatise dealing with the comet of 1577, since the second one, concerning the more modest comet of 1580, was clearly inferior, as Maestlin himself admitted in his dedicatory letter to Duke Ludwig of Württemberg (1554–1593). Regarding the reason for this difference, Maestlin referred to the disturbances caused by his move to Heidelberg after he was appointed professor at the university, which coincided with the appearance of the comet in the sky.[43] Jarrell indicates that, "like the *Consideratio & Demonstratio* [the treatise of 1581], the *De cometa anni 1618* fails to live up to the 1578 work. There is no demonstration of an orbit, but there is a lengthy attack upon Peripatetic cometary theory".[44] This is true, but this mathematico–astronomical shortcoming of the treatise of 1618 and Maestlin's focusing, forty years later, on the more physical aspects of cometary theory are closely linked to the cultural context of the moment as well as to the origin of the new treatise, that is, the fact that it was commissioned by the Duke of Württemberg. For a better understanding of this difference in outlook, it is worth reconsidering briefly the aims and context of the cometary treatises of 1578 and 1581 (as made clear in the former).

Already in his dedicatory letter to the Duke—dated the fifth Sunday of Lent 1578 (*"Dominica Iudica"*; 16 March)—Maestlin asserts that, although destined for theology, the Duke's generosity has allowed him to devote his free time to studies in the Faculty of Arts, especially mathematics rather than philosophy.[45] After completing his studies at Tübingen, he served as a church deacon before he was appointed Professor of Mathematics at Heidelberg in 1580.[46] For this reason, his approach to the comet of 1577, as with the nova of 1572, was purely astronomico-mathematical: that is, its aim was to elaborate on the quantitative aspects relating to the celestial location and motion of the comet, including its distance from the earth. This had also been the approach adopted by Brahe, in his studies of both the nova and the comet. Also like Brahe, Maestlin refused to enter into astrological considerations and conjectures, since they lacked the

43 See Maestlin (1581, A3[v]): "Sed huic instituto meo, ne videlicet motus huius calculum integrum enodarem, moras iniecit loci conditionisque meae mutatio".

44 Jarrell (1975, 16).

45 Maestlin (1578, A3[v]): "Cels. Vestrae liberalitate et stipendiis eo deveni, ut in ocio potuerim incumbere, non tantum philosophicis artibus et scientiis, maxime vero iis, quae Mathemata redolent".

46 *Ibidem*: "verum etiam potuerim salvificae doctrinae sacrosanctae verbi divini, quantum Dei gratia mihi concessit, operam dare".

necessity and certainty of mathematical conclusions, even though he (still in the same spirit as Brahe) was fully conscious of the theological dimension of the phenomenon and advanced some pious conjectures about it.[47] Maestlin steadfastly refused to enter into astrological considerations, as he clearly stated later in his manuscripts on the new star of 1604[48] and the comet of 1618.[49] As a consequence, our analysis will focus on the distinctive feature of the first cometary treatise: its astronomico–mathematical approach and results.

As brilliantly shown by C.D. Hellman[50] and R.S. Westman,[51] Maestlin achieved an extremely accurate observation of the entire course of the comet even without elaborate instruments. He further applied to his observations the new geometrical device (introduced by Regiomontanus)[52] for calculating the daily parallax of the comet as an infallible method for determining its distance from the earth.[53] Maestlin concluded that, unlike earlier comets, the present one had to be in the heavens, since it lacked any sensible parallax.[54] Chapters 7 and 8 further specified that it was located in the *sphaera* of Venus, according to the heliocentric hypothesis and Copernican order of the planets.[55]

---

47 *Ibidem*, A 4^{r-v}: "Nam licet quae Astronomus de hoc Cometa dicere potest, compilaverim, quae tamen is portendat, ego coniecturas tantum, non ex Astrologiae fontibus promanantes, sed aliunde derivatas [from theology and Scripture] notavi. [...] Etsi enim hactenus Mathematicam abstractam et concretam mihi nonnihil familiarem fecerim, in concreta tamen, cui motuum coelestium considerationes subiacent, ego Astronomiae potius, quam Astrologiae incubui. [...] Quare iudicium Astrologicum mihi hic arrogare nec possum nec volo, sed id aliis relinquo, quorum multos video admodum esse solicitos, ut audacter (siquidem hoc facile est) divinent".

48 See Granada (2014, 95–96, 108).

49 See below, Chapter 9, where several passages from the rough copy, explicitly affirming his abstention from astrology, are quoted.

50 Hellman (1944, 137–159). Cf. also the excellent summary in Jarrell (1971, 108–122).

51 Westman (1972, 10–25).

52 In his *De Cometae magnitudine, longitudine ac de loco eius vero, problemata XVI*, first published by Johann Schöner in Nuremberg in 1531. Cf. the facsimile reproduction in Jervis, (1985, 170–193) and the study with translation (1985, 95–112).

53 Maestlin (1578, Chapter 3, 7–10). See also Hellman (1944, 135–137). On the problems with measuring parallax, see now Adam Mosley, "A Brief Note on Cometary Parallax", in Granada, Mosley, and Jardine (2014, 326–339).

54 Maestlin (1578, 10): "Cum itaque *universalis experientia, indubitatis Geometriae fundamentis superinstructa* evincat, Cometam iis passionibus, quibus Luna positus suos in horas variat, multo autem magis stella aliqua secunda inferior [a sublunary comet], destitui: *certissimum* evadit, eum longe supra sublunarem orbem evolasse, et in superiori aethereo mundo sedem sibi quaesivisse, ad cuius altitudinem terrae semidiameter, *quantum ad parallaxes, non habeat sensibilem proportionem*" (our italics).

55 Maestlin (1578 Chapter 7, 37–38): "Cum igitur indefessa meditatione usitatarum [that is, geocentric] hypothesium omnes sphaeras pertractarem et revolverem, nullam ex omni-

A few contemporary authors arrived at the same conclusion—with the exception of Copernicanism—through the same method of calculating parallax: Tycho Brahe (who published only in 1588), Cornelius Gemma, Wilhelm IV Landgrave of Hesse–Kassel (who published nothing) and Helisaeus Roeslin. However, the vast majority of observers attributed a sublunary position to the comet, according to Aristotle's cometary theory and the essential immutability of the heavens, incompatible with any kind of generation or corruption like that suggested by the appearance and disappearance of the comet of 1577.[56]

Maestlin strongly condemned any author who had concluded to the contrary.[57] He thus did not call into question his own conclusion, but argued that the 'sublunarists' had been misled by their inaccurate observations and had measured parallax wrongly. The important conclusion to be drawn, therefore, is that Maestlin (together with Brahe, Gemma, and observers of the nova of 1572, such as Thomas Digges and Jerónimo Muñoz) claimed for mathematical astronomers the right to establish, from mathematical premises, cosmological conclusions regarding the place of comets in the cosmos and even—just as Copernicus and Rheticus had done before him—the disposition and order of the world. Thus, the most significant contribution of Maestlin's tract was the very vindication of the cosmological conclusions made by mathematical astronomy, along with his parallel opposition to the pretensions of Aristotelian natural philosophers who meant to monopolise discussion on cosmological matters.[58]

This notwithstanding, Maestlin limited his affirmation to only the supralunar position of the present comet as an indubitable truth and did not question the sublunary position of the vast majority of previous comets. Though

---

bus inveni, quae qualicunque ratione Cometae huius motum salvaret. Sed cum necdum quiescere possem: iis sepositis, ad Copernici, post Ptolemaeum Astronomiae vere principis, mundi symmetriam me recepi. In ea examinatis universis, tandem lib. 6, cap. 2. ubi latitudines Veneris exponuntur, orbem quendam inveni, cuius magnitudinem et conversionem cum ad amussim apparentiis Cometae correspondere et satisfacere deprehendissem, sicut capite sequenti demonstrabitur: omnino mihi statuendum fuit, Cometam non in quoquam alio orbe, praeter hunc, in Veneris Sphaera sibi sedem elegisse". Cf. *ibidem*, 40, 54, 56 and Westman (1972, 16 ff.).

56 Besides Hellman (1944) see now Jerratsch (2020, 179–224) for the German *Flugschriften* on this comet.

57 In Chapter 4 ("Opiniones aliorum, qui hunc Cometam elementarem putant, recensentur et diluuntur"), Maestlin criticised Andreas Nolthius as well as two other unnamed authors. See Hellman (1944, 149–152).

58 Charlotte Methuen has rightly insisted on this. See Methuen (1998, 171–181).

more inclined towards the optical theory that interpreted comets as the product of solar illumination,[59] he did not reject, at least for sublunary comets, the Aristotelian theory of the generation of comets from terrestrial exhalations rising to the highest region of the air and there inflamed. Instead, Maestlin preferred to admit his doubt and declare the issue to be unresolved:

> Since we are also so uncertain about the matter of elementary comets, who shall explain their fire? Natural philosophers state that they are an inflammation of exhalations, others an illumination of them. No matter whom is given assent, *I remain in doubt*. That that matter, whatever it may be, is consumed, is clear from the fact that comets are not perpetual, but I dare not say whether a fiery flame is consumed or passes away by its very lack of stability. What shall be said, however, about this comet that shone in the high aether? Perhaps earthly vapours rise through the air and fire (if it does exist), through the sphere of the Moon, without being inflamed, or illuminated by the rays of the Sun, before they reach the sphere of Venus? Convinced by all this, I felt obliged to abandon the customary and commonly received opinion, seeing that it neither corresponds to this comet nor is consistent with itself. *But what may be the generation of comets, especially ethereal* (if generation should now be said rather than creation), *whether similar to elementary [generation] or different, I judge that nobody can show from philosophy* (our italics).[60]

This scepticism concerning the physical theory of comets, in contrast to the more certain mathematical conclusions regarding their distance and motion, gave way to a commitment to divine causality. In other words, Maestlin deemed

---

59 On this theory and its origin in the observation of the comets of the 1530s by Petrus Apianus and Girolamo Fracastoro, see Barker (1993), Kokott (1994) and Jerratsch (2020, 122–127).

60 Maestlin (1578, 18–19): "Cumque de materia Cometarum elementarium quoque tam incerti simus, quis ipsorum incendium exponet? Physici halituum incendium statuunt, alii eorundem illuminationem. Utris assentiendum sit, *haereo*. Materiam illam, quaecunque etiam sit, consumi, ex eo patet, quod Cometae perpetui non sunt, sed flamma ne ignea absumat, an vero ipsa sua instabilitate transeat, *non dixero*. Quid vero de hoc Cometa dicendum erit, qui in alto aethere illuxit? num per aërem, et (si quis est) ignem, sphaeram item Lunae, terreni vapores assurgunt, non inflammati, vel radiis Solis illustrati, priusquam in Veneris orbem [...] pervenerunt? His et ego convictus ab usitata et vulgo recepta opinione deficere coactus sum, videns, eam nec huic Cometae correspondere, nec etiam inter se cohaerere. *Qualis autem sit Cometarum, praesertim aethereorum, generatio* (si modo generatio potius, quam creatio dicenda sit) *elementari ne conformis aut dissimilis, ex Philosophica Schola neminem ostendere posse arbitror*".

the present celestial comet—extremely infrequent, if not unique in history—a divine, miraculous creation, just like the recent nova in Cassiopeia. Thus, Maestlin continued:

> But if, leaving it [philosophy] aside, he should return to the theatre of most holy Theology, he will draw out from its secrets if not all, at least enough to be content with when seeking after them. For he will hit upon a creator of the whole world and nature, endowed with so great a wisdom, virtue and power that he can do whatever he should will, to such an extent that the entire universe is not tied together with laws so tight that he cannot loosen them. [...] Who shall deny that he has reserved many other [wonders] in his omnipotent and still unweakened hand, some of which he has determined to exhibit to us, without unfolding them plainly? There is no doubt that the present comet belongs to this genre of things, since, in fact, it brilliantly ridicules the acute speculations of every philosopher as altogether false. Thus, I happily bow to the opinion of those who assert that comets must be restored among the secrets of nature, seeing as they are not the hidden works of nature, but of God.[61]

Maestlin completed his analysis by neutralising any potential opposition from Aristotle, arguing that the Stagirite might have only known of aerial comets and, most importantly, was unaware of the mathematical doctrine of parallax. Had he known it, Maestlin argued, Aristotle would have certainly changed his mind about the comet of 1577, as well as his erroneous definition of the Milky Way as a sublunary phenomenon:

> No doubt, if Aristotle had been sure about ethereal comets found by parallax (for those who placed comets in the heavens before him were unable

61 Maestlin (1578, 19): "Sed si ea relicta in S.S. Theologiae theatrum se receperit, etsi non universa, talia tamen, quibus haec inquirens contentus esse potest, ex eius secretis depromet. Nam in ea offendet, totius mundi et naturae opificem tanta sapientia, virtute et potentia praeditum, ut quaecunque voluerit, possit, adeo ut tam strictis legibus totum universum colligatum non sit, quin ipse solvere valeat. [...] Quis negabit, eum plurima alia omnipotenti suae manui, quae nondum infirmata est, reservasse, quorum quaedam duntaxat nobis monstrare, non autem dilucide explicare voluit? Ex horum genere certe praesens Cometa quod sit, dubium non est: siquidem ipse cunctas omnium Philosophorum argutas speculationes false et mirifice ludificat. Itaque lubens descendo in eorum sententiam, qui Cometas in Naturae secretis reponendos, et opera occulta non Naturae, sed Dei esse asserunt". Maestlin seems to refer to Thaddaeus Hagecius's work on the nova of 1572. Cf. Hagecius (1574, p. 55): "Cometas in secretis naturae reponendos, esseque occulta opera non naturae sed Dei: ac eos ostendi ad tempus terris, ut essent nuntii futurorum".

to demonstrate what they claimed), he certainly would have changed his preconceived opinion. However, here, as in the description of the Milky Way, he deserves indulgence due to the harshness of the times.[62]

5 From 1578 to 1618

Maestlin's interpretation of the comet of 1577 was exceptional. Leaving aside the four other authors mentioned above, the vast majority of astronomers and astrologers remained faithful to Aristotle and the traditional notion of the changeless heavens. Even the dissolution of the solid celestial spheres had to wait several years before coming under discussion among astronomers.[63] Needless to say, this was also the standpoint of the majority of natural philosophers, who continued to follow Aristotle faithfully. As a consequence, the teaching of natural philosophy, including comets, remained essentially unaffected by the comet of 1577. The fact that otherwise isolated philosophers such as Giordano Bruno and Bernardino Telesio[64] adopted the new concept of comets was similarly an exception (and pertained to very different cosmological perspectives).

The appearance of further comets in the following years up to 1607 and the invention of the telescope gave occasion for innovators to reassert and amplify their own theories. Some affirmed that *all* comets were celestial phenomena and, consequently, that they were formed or generated from celestial matter. At the same time, the frequent appearance of celestial comets allowed for a natural explanation involving secondary causes, without excluding a connexion with divine design and purposes—especially among Protestant authors, who were often more sensitive to the eschatological import of celestial novel-

62 Maestlin (1578, 17): "Dubium non est, si Aristoteles de aethereis per parallaxes inventis certus fuisset (qui enim ante ipsum Cometas in coelo reponebant, quae dicebant, probare non poterant) profecto sententiam conceptam mutasset. Sed hic, ut et in descriptione Galaxiae, propter temporum iniuriam veniam meretur". Similarly, Habrecht (1618, 9) affirms that "er [Aristotle] entweder der Mathematischen Instrumenten gebrauch nicht gehabt oder wie noch heute etliche thun [in the margin: Keckermannus; on this Aristotelian philosopher, see Chapter 6 below] denselben verachtet, hat Er hierinnen sehr weit gefehlet. Dann erstlich hat er auß ubersehung der *parallaxeon*, die *galaxiam*, oder weisse Milch, oder Jacobstraß under den Mon gesetzet, welche doch in dem obern Firmament, bey den Fixensternen nothwendig sein muß, dieweil sie kein *parallaxin* im wenigsten nicht hat".

63 On this subject, see Granada (2006).

64 For Bruno, see Tessicini (2010); for Telesio, see Telesio (2012).

ties. This was the case for Tycho Brahe and Christoph Rothmann,[65] as well as for Maestlin.[66] After writing a second treatise on the comet of 1580 in which he reached the same conclusions as in 1578, the Tübingen astronomer restated the matter in his manuscript treatise on the nova of 1604 in the following terms:

> The school of the Peripatetics has taught us thus far that all meteors are born [and] arise in the elemental region, that is, the superior region of the air, and some think that this opinion must be retained at all costs. For this reason, some of them have attempted with all their might and power of mind to remove the star of 1572 and the comets of preceding years to this same region. But many of them, after they saw themselves, refuted by the observations of the mathematical experts, that the star of 1572 and the comets of 1577, 1580, 1582, 1585, 1590, 1593 and 1596 clearly lacked all perceptible parallax, and for this reason, in keeping with the doctrine of parallax (which alone in the measurement of distances, without contradiction or exception, can neither deceive nor be deceived, and so without contradiction must be recognised as the rule of truth), it is completely certain that none of them could have arisen in the elemental region, but all of them [arose] in the aethereal region of the heavens, and indeed far beyond the Moon and all of them beyond those orbs in relation to which the earth has some sensible proportion of magnitude: likewise, after they learned from genuine foundations that either they themselves or certain others thought from their own calculations that they had found or distorted some parallax, they had erred far from this doctrine: finally, they began to become less hostile and having become more equitable to the

65 In a single work (his manuscript treatise on the comet of 1585, written between 1585 and 1589) Rothmann turned from explaining the comet as an illumination of terrestrial exhalations elevated to the heavens to conceiving it as generated in the heavens from celestial matter. See Granada, Mosley, and Jardine (2014, 30–36).

66 Maestlin never discussed comets in his *Epitome astronomiae* (several editions between 1582 and 1624), an astronomy handbook that embraced the traditional hypotheses and conception of this discipline. Thus, in this work Maestlin always conceded that the heavens underwent no changes with the exception of circular motion. Cf. Maestlin (1610, 34): "[Aetherea regio] est ea pars mundi, quae est supra sphaerae Lunae [...] continens omnium syderum orbes: lucida, simplicissima, *omnis motus, praeterquam rotundi expers*" (our italics). See also Methuen (1996, 236, 238–240). Methuen rightly affirms that Maestlin gave his more advanced students the opportunity, in the theses he proposed for the university disputations, to defend more innovative positions. As is known, Maestlin fully endorsed the a priori confirmation of Copernican cosmology by Kepler through the polyhedral hypothesis in the *Mysterium cosmographicum* (1596), in whose publication Maestlin played a decisive role.

> aethereal region, the power which they foolishly but insistently denied to it, as if nothing new could arise or happen there, they restored, confessing that in investigating the truth, it is not what reason persuades, nor common opinion (which, unless it relies upon firm foundations is oftentimes no different from the fallacious opinion of the ignorant multitude), but what Experience, as the proper measure of rational calculations, transmits that must be considered. It is not that anyone should believe that in this way the old arts and sciences are to be destroyed, as if proceeding on this path a new one is finally being sought by us. But the true and ancient sciences are not being unsettled, much less are they being overturned; however, the erroneous opinions in them, having been defended as true up until now, are pulled out and removed, and in their place is planted the indubitable Truth. Moreover, by the common chain of Truth all doctrines are elaborated and accepted, and sciences, the more eminent as well as the more skilful, are connected, for all of them are established and come from God, who is the truth itself.[67]

Here, in 1605, there is some reason to believe that Maestlin was optimistic in his evaluation of the evolution and progress being made by natural philosophers, not to mention astronomers. That is to say, he was expressing his hope for the eventual advance of traditional astronomers and natural philosophers towards accepting the celestial location of comets and the mutability of the heavens. At the same time, Maestlin tried to persuade traditionalists that the new cometary theory did not imply a revolution in science or the destruction of the "old arts and sciences" (namely Aristotelian philosophy), but merely the replacement of erroneous particular conceptions with new, more accurate ones.

However, the situation remained relatively unchanged and the traditional Peripatetic views in the domain of natural philosophy and university teaching continued to hold sway.[68] Maestlin himself was to perceive this shortly afterwards, as we may infer from his treatise on the comet of 1618. Here, he took note of the conservative reaction by an influential Aristotelian German philosopher, Bartholomaeus Keckermann (ca. 1572–1609) who, in his public instruction as well as his several dissertations and works published immediately before his death and repeatedly reprinted afterwards, vehemently attacked innovations in cosmology and cometary theory. Keckermann passionately supported the traditional notion of a naturally immutable heaven (susceptible only to new

67 Granada (2014, 120–121), as translated by Patrick J. Boner. For the original Latin, see *ibidem*, 114.

68 On the sixteenth-century meteorological tradition, see Vermij (2010) and Martin (2011).

creation by virtue of divine omnipotence) and of Aristotelian cometary theory. Since Keckermann's views were the explicit target of Maestlin's criticism in his treatise on the comet of 1618, it is worth examining Keckermann's claims, all the more so since they provide us with a clue for better understanding the particularity of Maestlin's argument in 1619 and the differences in his position vis-à-vis 1578—that is, the less mathematical and more philosophical and physical character of his discourse on the later comet.

6 Bartholomaeus Keckermann and His Assault on Celestial Comets

After distinguishing between the second and third comets in the first pages (pp. 9–11) of the clean copy belonging to the second chapter, Maestlin devotes the third chapter (pp. 11–15) to the motion of the third comet. In particular, Maestlin examines the path of the comet among the constellations and its variation in speed, as well as the various characteristics of its tail. In the fourth chapter (pp. 15–17), he turns to the contemporary conflict in philosophy between the majority ("die mehrerthail *Philosophi*"), who follow the doctrine of Aristotle—namely, that comets are the sublunary result of dry and warm exhalations inflamed in the upper region of the air, their short duration excluding them from the immutable region of the aethereal heavens—and "the others" (with no indication of number or names given; Maestlin refers only to "some philosophers"). From the circumstances of cometary motion (proportional to a certain degree; "gewissen proportionierten fortgang"), the latter defend the celestial nature and location of comets, considering them a kind of star, eternally present in the heavens though seldom visible.[69]

In this unbalanced confrontation, it is not clear who might belong to the minority of philosophers. Maestlin may have in mind classical authors such as Seneca (or those criticised by Aristotle in *Meteorologica* I, 6) or near contemporaries like Cardano.[70] At any rate, among the majority of Aristotelian

69 Maestlin (1619, 17): "Daher ettliche *Philosophi* haben die Cometen nicht nur in den himmel erhebt, sonder sie gar under die Zahl der planeten gesetzt, und darfür gehalten, das die Cometen seien gewisse Stern, so am himmel stätigs wie andere Stern stehen, aber bleiben under der Sonnen hellen schein verborgen, bis ir aigner *motus* sie in gewissen zeiten laße, gleich wie zu zeitten der *Mercurius* thut, herfür lauffen".

70 See Cardano (2004, 364): "verius est, coelum esse sideribus pluribus, sed non admodum densis, plenum; quod cum aer siccescit et attenuatur, vel etiam aliis ex causis, oculis nostris se subiicit". We may exclude the possibility that Maestlin knew Giordano Bruno's *De immenso et innumerabilibus*.

detractors of celestial comets were some who did not refrain from denouncing the opposing standpoint as a "mortal sin, for not only would natural philosophy but the entire field of philosophy come completely crashing down from it".[71] Maestlin is most probably referring here to Keckermann, whom he introduces explicitly and refutes at length in Chapter 8.

Bartholomaeus Keckermann (ca. 1572–1609) was, from 1602 until his untimely death, Professor of Philosophy at the Gymnasium in his native city of Danzig (now Gdansk).[72] There, he wrote many works covering the full extent of philosophy. These works reached a wide audience in the first half of the seventeenth century and, due to their scholarly character, exercised an important influence on the philosophical training of younger generations. Among these works, two deal with comets in general and with the recent—supposedly celestial—novelties in particular: the first, published for the first time in 1606 and reissued in 1611, was titled *Theoremata exegetica De cometis, in genere: Et in specie, de tribus illis mirabilibus Facibus, quae Anno 1572. & denique Anno praeterito 1604 apparuerunt*. This was a set of theses for a disputation held in May 1605 (with Peter Crüger (1580–1639), later Professor of Mathematics and Poetry in Danzig and an avid supralunarist, who served as a respondent), to which was appended a *Diaskepsis De observationibus cometarum per instrumenta Astronomica, atque adeo etiam eius, qui apparuit Anno praecedenti 1604, Octobri, Novembri et Decembri mensibus*.[73] The second work was titled *Systema Physicum septem libris adornatum, Et Anno Christi* MDCVII. *publice propositum*, whose sixth book deals with *De meteoris* and features in the fifth chapter a long discussion *De cometis*.[74] Maestlin explicitly mentions both works[75] and builds upon them in his refutation of the still predominant concept of comets.

71 Maestlin (1619, 17): "Ettliche darfür halten, es seye diser [the Aristotelian doctrine of comets] zu widersprechen, gleichsam ein schwehre Todsünd, dann hiedurch würde nit nur die *Physica Scientia*, sonder die gantze *philosophia* samentlich zu grund und boden fallen".

72 For a first approach to Keckermann, see Freedman (1997), who, however, does not contemplate our current subject. See also Jensen (2006, 36–50) and Granada (2019).

73 These were printed in Keckermann (1606). For the *Theoremata exegetica De cometis*, see 344–376; for the *Diaskepsis*, see 377–417. On Crüger and his own work on the comet of 1618, see Granada (2016, 53–56), Jerratsch (2020, 349–383).

74 Keckermann (1610). *Systema Physicum septem libris adornatum*, Hanau 1610, reissued in 1612, 1617 and 1623. *De meteoris* extends over pp. 624–835 in the 1623 edition (printed by Petrus Antonius) that we have used. The chapter *De cometis* is included on pp. 669–785.

75 Maestlin (1619, 41): "Aber alhie wehret sich mit allen kräfften neben ettlich andern, in sonderhait Bartholomaeus Keckermann, in *Systemate Physico*, und in *Disputationibus Physicis, Disput. extraord. de Cometis, et adiuncta* διασκέψει".

Keckermann strongly defends the Aristotelian theory of comets (*inflammations* of *earthly exhalations* in the upper region of the *air*)[76] as uniquely compatible with the immutability of the aethereal heavens and with the most certain doctrine of the heterogeneity between the sublunary and supralunary worlds. This fiercely conservative outlook, however, does not mean that Keckermann was insufficiently informed about contemporary claims in cosmology and, in particular, the celestial novelties since 1572. Moreover, Keckermann greatly admired Tycho Brahe, whose works on the nova of Cassiopeia and the comet of 1577, with the discussion of contemporary literature they contained, he had studied closely. Keckermann also mentions Maestlin as one of the authors on the nova of 1572 discussed by Brahe in his *Progymnasmata*, so we can confidently conclude that he was at least familiar with Maestlin's interpretation of the nova.[77]

Keckermann knew perfectly well that the strength of the innovators resided in the presumed accuracy of their observations and in the necessary conclusions derived from the measurement of parallax. He was also well aware that the celestial location of comets (and novas, according to Brahe and Maestlin), intended as natural phenomena, wholly undermined Aristotelian cosmology[78] and was convinced that it implied the destruction of natural science as taught in the schools:

> Concerning the place and location, as well as the distance of this Torch [nova of 1604] from the Earth, an anxious and apprehensive meditation

76 Keckermann (1606), theorems 10–12: "Omnino naturalis est causa cometarum", "Non tamen in aetherea regione generantur, ac proinde nec stellae sunt, nec stellarum configurationes", "Relinquitur ergo, cometas esse corpora sublunaria, et quidem exhalationes calidas et viscosas, virtute astrorum in supremam regionis aeris extractas, ibidemque non a Sole κατ' ἀνάκλασιν illustratas, sed revera accensas et flagrantes", pp. 348f. Keckermann also strongly rejects the optical theory of Apian, Gemma Frisius and Cardano. Gindhart (2006, 249–255) offers an excellent summary of Keckermann's views on our subject, without, however, considering Maestlin.

77 Keckermann (1606, 370): "Nonnulli, quorum Coryphaeus Tycho Brahe, ille omnium aevi nostri Mathematicorum princeps, hoc sidus nequaquam in elementari mundo, sed longe supra lunam et omnes alios Planetas in ipso stellarum fixarum orbe efulsisse tenaciter demonstrare conantur. Atque huius classis sunt Thaddaeus Hagecius Bohemus, Paulus Fabricius, Bartholomaeus Reisacherus [...] Michael Moestlinus".

78 Keckermann (1606, 693–694): "Valde periculosa est ista sententia, utpote quae non parum labefactet totam doctrinam de ordine ac distinctione partium mundi, deque distinctione corporis coelestis a corporibus elementaribus. Si enim in ipsam usque substantiam coeli fumus aliquis sulphureus delatus fuit, ibique incensus, utique ipsa coeli substantia mutata et alterata fuit".

> torments me; and all the more since I see that in the last 33 years observations of some comets have, from the very same foundations, led to the overthrow of the principal part of natural science so well established in the schools regarding the immutability, constancy and purity of the celestial body (so different in nature from the remaining lower bodies), as well as the doctrine concerning the element of fire, air and, at the same time, almost the whole of Meteorology.[79]

To forestall this painful development, Keckermann endeavoured to discredit both the astronomical observations and the parallactic demonstrations. The former, he argued, lacked credibility for three main reasons: 1) the inaccuracy of the observational instruments themselves;[80] 2) the refraction of the rays coming from the cometary phenomena when they passed through fire and air, the result being that "those celestial bodies cannot be observed in their proper places";[81] and 3) the eccentricity of our place of observation on the earth's surface.[82] As a consequence, the geometrical determination of distance by parallactic measurement could be neither accurate nor precise:

> However, if they could not be observed correctly through astronomical instruments, assuredly nothing certain or proper can be pronounced concerning the parallax of comets. Yet it is in the observation of their parallax

79 *Diaskepsis* in Keckermann (1606, 378–379): "De loco tamen et situ, atque adeo distantia istius Facis [nova of 1604] a terra, sollicita me sane et anxia torquet cogitatio; idque eo magis, quod videam, ab annis nunc 33. cometarum aliquot observationes eo pertinuisse, ut bene constituta in Scholis scientiae naturalis pars princeps de caelestis corporis immutabilitate, constantia, puritate, atque adeo a reliquis corporibus inferioribus naturae distinctione, itemque doctrina de elemento ignis, aeris, simulque universa pene Metereologia ex ipsis usque fundamentis labefactetur". Cf. also 380: "[...] unde verendum fuerit, ne pro Systemate scientiae naturalis, opinionum variarum ac dubiarum Chaos in scholis simus habituri".

80 *Diaskepsis* in Keckermann (1606, 382): "Haec ergo sententiae meae summa est: Nullius omnino proprie et vere dicti cometae situm et motum per instrumenta Astronomica sic observari potuisse hactenus, aut posse adhuc, ut ex observatione ista concludi firmiter et demonstrari queat, cometalem ullum fumum ad aethera usque evectum esse, ibique accensum".

81 *Diaskepsis* in Keckermann (1606, 383): "idcirco sidera non posse in propriis et veris locis per instrumenta observari".

82 *Diaskepsis* in Keckermann (1606, 384): "neque enim videri possit erroris expers observatio, quae centrum mundi cum centro instrumenti confundit; exactissima nempe observatio, ea tantum futura est, quae fit per instrumentum et oculum observatoris, collocatum in ipso mundi centro, utpote cum ex caelo, tamquam ex circumferentia radii caelestes, tamquam lineae, ad ipsum istius circuli centrum convenienter terminentur".

> that the first foundation of their observation, as it were, is established, whereby comets are held to be generated and to have adhered in the very same ethereal region.[83]

For all these reasons, observing comets could produce an error of "many German miles" in determining their distance.[84] Nevertheless, in order to avoid the total discrediting of astronomy, Keckermann declared himself ready to accept that this criticism did not affect the observation of true heavenly bodies (stars and planets), but only the 'lights' (*faces*) seen from 1572, which, being only comets, were not primary celestial bodies, perfectly round and moving uniformly.[85]

The many and profound discrepancies between astronomers in their observations and parallactic measurements of these phenomena—some placed them in the stellar sphere, others in the planetary region, and still others in the sublunary region—encouraged Keckermann to consider the parallactic method as unreliable when applied to comets:

> No doubt heavenly bodies are observable with mathematical instruments: when it comes to comets, however, favor is on our side if, persuaded thus far by many and great reasons, we should then believe that they can never be rightly observed, even by those highly sought out and great instruments of Tycho Brahe.[86]

Keckermann did not even contemplate the possibility that comets might be celestial bodies; there were too many discrepancies between observers due to their different degrees of observational accuracy and varying measure of competence as mathematicians. On the contrary, he took the immutability of the heavens for granted and concluded from it that comets were necessarily nat-

83 *Diaskepsis* in Keckermann (1606, 399): "Quod si autem observari recte per instrumenta astronomica non potuerunt, utique nec de eorum [comets] parallaxi certi quicquam et indubitati potest pronunciari, in cuius tamen parallaxeos observatione primum quasi fundamentum positum est eius observationis, qua feruntur cometae in ipsa aetherea regione generati esse et haesisse". Cf. *Systema physicum*, in Keckermann (1623, 719–720).

84 *Diaskepsis* in Keckermann (1606, 389).

85 *Diaskepsis* in Keckermann (1606, 385 ff.).

86 *Diaskepsis* in Keckermann (1606, 394): "Nimirum corpora coelestia sunt instrumentis Mathematicis observabilia: De cometis autem venia nobis datur, si multis et magnis rationibus adhuc persuasi ita credamus, eos nequidem per Tychonis Brahei illa tam exquisita tamque ampla instrumenta observari unquam recte potuisse". Cf. *Systema physicum*, in Keckermann (1623, 720–721).

ural phenomena in the sublunary region and therefore unsuitable objects for astronomical study. The differences between astronomers were thus seen as simply the result of applying to comets a method suitable for a completely different class of objects. The Danzig philosopher could only accept the celestial position of the phenomena observed between 1572 and 1604 as *extraordinary* or *supernatural* products of God's omnipotence, intended to warn mankind of His wrath and perhaps of the impending end of the world as well. But in this case, these celestial phenomena did not overthrow the Aristotelian doctrine or physical science in general, since both applied only to the *natural* world order. Therefore, the phenomena of 1572 and 1604 could be new stars, just as Brahe—and Maestlin, we should add—supposed. To distinguish these divine miracles from the natural course of events, Keckermann states:

> I would tolerate their opinion and not fight rashly against it, insofar as it neither overturns nor undermines meteorological doctrine or any other part of physical science. For it does not follow that, [since] new stars have appeared in the heavens, therefore the doctrine transmitted by Aristotle and other natural philosophers concerning comets and their generation, as well as the purity and perpetuity of celestial bodies, can no longer be maintained as true. New bodies may indeed be produced in the heavens by the miraculous power of God, without withdrawing any of the ordinary principles, forces or ways of acting in nature. [...] Certainly Aristotle never denied that the first cause can act without secondary causes, or in a different way than the secondary; it has never been denied that God can produce something new. It does not follow that, [since] *God often acts extraordinarily, apart from nature, he does not therefore act ordinarily with nature*. Nor does it then follow that, [since] *God often makes miracles, therefore the foundations of natural science and the principles of celestial nature, and of the generation of meteors, are overthrown*. Miracles are beyond nature, but they do not oppose or overturn nature.[87]

---

87 *Diaskepsis* in Keckermann (1606, 399–400): "eorum ego sententiam et tolerarim libentius, et non impugnarim temere, utpote quae nec doctrinam Meteorologicam, nec ullam aliam scientiae Physicae partem labefactat aut evertit. Neque enim consequens est: Apparuerunt in caelo novae stellae; *Ergo*, doctrina, quae ab Aristotele, et aliis Physicis tradita est de cometis et eorum generatione, itemque de caelestis corporis puritate et perpetuitate, veritatem suam amplius tueri nequit. Possunt namque in caelo nova corpora produci per miraculosam Dei potentiam, ut tamen nihil interea ordinariis naturae principiis, viribus, et agendi modis decedat. [...] Nusquam sane negavit Aristoteles, primam causam posse agere sine secundis, et aliter quam secundas; nusquam inficiatus est, Deum posse aliquid novi producere; nec sequitur, *Deus saepe absque natura agit extraordinarie, ergo cum*

Keckermann concludes with an appeal to re-establish concord between astronomy and physical science, a continuity that had been deplorably destroyed, with ensuing disaster, by conflicting interpretations of the recent extraordinary phenomena.[88] The restoration of concord and the integrity of natural science was not to be gained by completely subverting the inherited science (that is, by means of a revolution in science). It sufficed, Keckermann believed, that astronomers (and thus Maestlin) should cease to deal with comets and limit themselves to their proper field of celestial phenomena, conceding the study of comets to their appropriate owner, the *physici*: "since comets are not astronomical, but physical bodies, the astronomer, as an astronomer, will not lawfully deal with comets, but the natural philosopher".[89] Otherwise, as Keckermann had previously threatened, such a subversive doctrine of mathematician astronomers mingling with physics should be banished from the schools:

> For this reason, I strongly prefer that this opinion of fumes carried up to the very aether, mingled with that most pure celestial body, and set on fire in the very same place, would either be plainly ignored or spread only among the schools of philosophers with this solemn formula: IT IS NOT EVIDENT; for it is not as easy to restore good disciplines as it is to bring them down.[90]

natura non agit ordinarie; Nec denique sequitur, *Deus saepe facit miracula, ergo evertuntur Physicae scientiae principia et praecepta de natura caeli, deque generatione Meteororum.* Sunt supra naturam miracula, nec tamen naturam oppugnant aut evertunt". Cf. *Systema physicum*, in Keckermann (1623, 768–769).

88 *Diaskepsis* in Keckermann (1606, 414–415), where Keckermann expresses his affliction: "dolorem meum, quo me sentio affici, cum animadverto [...] viam sterni ad committendas inter se duas illas plane sororia necessitudine devinctas scientias, Physicam et Astronomiam [...]; venia detur iusto dolori, et meo pro naturalis scientiae integritate sive zelo, sive voto".

89 *Diaskepsis* in Keckermann (1606, 417): "neque enim Astronomica corpora Cometae sunt, sed Physica; nec de cometis Astronomus, qua Astronomus, legitime tractabit, sed physicus".

90 *Diaskepsis* in Keckermann (1606, 381): "Quocirca vehementer optem, istam de fumis ad ipsum usque aethera evectis, cumque purissimo illo corpore caelesti confusis, ibidemque accensis, opinionem vel non audiri plane, vel ampliari saltem in Philosophorum scholis cum illa solenni formula: NON LIQUET; neque enim tam facile restaurantur bonae disciplinae, quam convelluntur".

7 Maestlin's Reply to Keckermann and the Partial Preservation of Aristotle

Maestlin's reply to Keckermann appears in Chapter 8, after he has prepared the ground for it in the previous chapters. This time, his discourse is not merely mathematical; at least, it is less mathematical than those of 1578 and 1580. Maestlin now addresses the issues under discussion from a more philosophical or more meteorological point of view, confronting the old Aristotelian cometary theory with new results. In addition, he expands the debate to the novelties recently discovered by the telescope on the surface of the Moon and on the Sun, that is, the so-called sunspots. Maestlin's strategy runs parallel to that of a contemporary publication by the younger professor of mathematics at Leipzig, Philip Müller.[91] Müller's treatise, entitled *De Cometa Anni* M.DC.XVIII. *Commentatio PhysicoMathematica specialis & generalis*, also reacted strongly to the criticism of Keckermann.[92] Lacking instruments and confessing his insuf-

---

91 See Gindhart (2006, 234–243).

92 Cf. Müller (1619, 14): "Neque enim audiendus est caeteroquin perspicax, sed in hoc minus oculatus quidam magni Philosophus nominis (Keckermannus in disputat. quod extraordin.) qui propterea fidem derogat observatis Astronomorum de Cometae situ, loco, distantia a terra, magnitudine, etc.". Müller refers to Keckermann's extraordinary disputation held in May 1605 with Peter Crüger as respondent. Keckermann's criticism reached wide notoriety in Germany, since it was also known and vigorously rejected, on account of the comet of 1618, in Danzig by his former disciple Peter Crüger (in his *Uranodromus Cometicus*, Danzig 1619; Chapters 16 and 17, pp. 98–113, are devoted to Keckermann) and in Alsace by Isaac Habrecht. Cf. Crüger (1619, 112 f.): "Vir hic [Keckermann] tantus est, ut eius autoritate quidam tanquam oraculo nitantur. Ipse vero [...] passim eo laborat, ut Aristotelem nusquam errasse docere possit. [...] Ego tam Aristotelem quam Keckermannum ex animo veneror, sed ita tamen ut utrumque Hominem fuisse nec unos omnia potuisse sentiam"); Habrecht (1618, 16): "niemand zagen und klagen darf, wie der Kleinmütige Keckermann, die gantze Natur muß undergehn, wo man im geringsten ein änderung in den Himmel wolte bringen"; 51 f.: "Denjenigen aber die solche Observationen mit geometrischen Instrumenten und Nachrechnungen ganz verachten, soll man darmit etlichermassen die Mäuler stopffen, dann anderer antwort seind sie als blinde Polyphemen und unverständige Cyclopen kaum werth. Der fürtreffliche Philosophus Keckermann hat in einer Disputation von den Cometen auch solche Mathematische Observation [of parallax; in the lines just above, Habrecht expresses his enthusiastic praise, in a true Melanchthonian spirit, of Arithmetic and Geometry as the wings enabling us to rise to the heavens] ganz und gar verachtet und verlachet, das an einem so gelehrten Mann hoch zu verwundern". On Peter Crüger's reply to Keckermann, see Granada (2016, 287–290) and Jerratsch (2020, 367–371). Not very far from Danzig, in Greifswald, Johann Döling published in 1619 a *Discursus Mathematico-Physicus in quo De Cometis contra Aristotelicos, imprimis vero Bartholomaeum Keckermannum Dantiscanum pro Nobilissimo Tychone Braheo, nostro seculi Atlante* [...] *disseritur*, whose title is fully indicative of Keckermann's role in the discussion on comets ten years

ficient mathematical ability for the task, Müller appealed, nevertheless, to the observations of more skilled mathematicians, including Maestlin, who is cited explicitly for his parallactic conclusions regarding the comets of 1577 and 1580.[93] Müller called on the above mathematicians to demonstrate definitively the supralunarity of comets through the accurate measurement of parallax. In Chapter 6 (entitled "Digressio pro asserenda certitudine mathematicorum observationum"), Müller declares:

> From them [Mathematicians], I eagerly anticipate an exact measure of parallax for this comet, so that through it a settlement is finally achieved among the schools on the nature of comets and this most noble controversy, plagued by so much exasperation of spirits on both sides and with such protracted zeal among the litigants, does not remain forever undecided. For parallax is the most secure means to guiding us rightly and unfailingly to knowledge of a matter so abstruse and far removed from the senses.[94]

For his own part, Maestlin states in Chapter 5 of his treatise that the contrasting opinions of the philosophers concerning the celestial or elementary nature of comets are, in both cases, simply worthless 'conjectures'.[95] Up to that point in time, both parties were thought to have adduced only 'prejudices' and not a single solid demonstration. The reason for this failure was that both sides had sought to provide a causal explanation of comets (in the field of *dioti*, that

after his death. On this intervention, derived from a public disputation and consisting of 172 theses directed in large part against Keckermann, see Gindhart (2006, *ad indicem*). Last but not least, another important critic of Keckermann was Christophorus Hunichius (d. 1623), a professor at Stettin's *Paedagogium illustre*, who entered into controversy with him from 1606. On this interesting figure, significant also for his position on Copernicanism, see Granada (2019, 221–232).

93 Müller (1619, 34).

94 *Ibidem*, 13–14: "Ab iisdem [Mathematicians] avide expecto exactam Parallaxeos rationem in hoc Cometa, ut per eam transigatur tandem in scholis super natura Cometarum, neque controversia haec nobilissima, tanta utrinque animorum exacerbatione, tam diuturnis litigantium studiis vexata maneat perpetuo indecisa. Est enim Parallaxis tutissimum medium, quod in rei tam abstrusae, tam longe remotae a sensibus cognitionem, recta nos et infallibiliter perducat". Cf. also pp. 33–34, 43–44.

95 Maestlin (1619, 18): "Es werden in diser disputation zu beeden Seitten, wie vermeldet worden, *argumenta* auf die Baan gebracht. Aber wan man selbige alle [...] erwigt, so seindts nichts, als nur *coniecturae* und muttmaßungen, welche im grund nichts gewißes beweisen, sonder es sind bis daher *argumenta argumentis, ratiocinationes ratiocinationibus, coniecturae coniecturis*, eines dem andern entgegen gesetzt, welche zuletzt samentlich sovil als **nichts** gelten".

is, the perfect causal demonstration *propter causam*, according to Aristotelian logic),[96] without first establishing the fact (*oti*), in this case the precise location of the comet or its distance from the earth.[97]

According to Maestlin, this demonstration of the fact (the distance) is only provided by mathematics, that is, by the *Doctrina Triangulorum* (trigonometry), whose application to comets yields their height with the same necessity and certainty as mathematicians here on earth are able to determine the height of a tower. However, in contrast to the very technical tone of the treatise of 1578 (and even that of 1580), Maestlin now adopts a more pedagogical and persuasive approach, very probably in accordance with the character of the treatise as a report commissioned by the Duke, and first presents the widely known example of the mathematical demonstration of the height of a tower resulting from the parallactic angle taken from two known points equally distant from the tower. With the aid of a second example and a second illustration,[98] Maestlin clearly shows that the problem of cometary distance is identical, and therefore can be solved, through mathematics, with the same absolute certainty as the distance of any object on earth. Moreover, Maestlin adds, when the distance of the comet increases, the parallactic angle with respect to two points on the earth (the centre of the earth and the point of observation on its surface) diminishes. As a result, at a very long distance no parallactic angle, and thus no parallax, becomes observable. Since the parallax of the Moon amounts to approximately one degree, the absence of parallax in a comet clearly demonstrates that it is more distant than the Moon and necessarily located in the aethereal heavens. This is the 'magic', Maestlin concludes, that astronomers employ to ascend to the heavens and conclude with mathematical certitude (*gewiß*) whether a comet is above or below the Moon:

> These are the songs and spells (as the poets say) by which the Moon and other stars are drawn down from the sky, and these are the ladders by which the astronomers ascend to the heavens and measure the heights of

---

96 Maestlin himself refers (*ibidem*) to Aristotle, *Analytica posteriora*, II, 2 and 8.

97 *Ibidem*: "Thus, in our present matter one should and must begin by setting aside all conjectures and determine from the right and proper reason where this or that comet may be, whether a comet has appeared in the air or in the heavens (whatever its origin may be and should it follow a regular motion or not), and that one could know for certain such things *a priori* and by demonstration". On this distinction (corresponding to the Latin terms *demonstratio quia* and *demonstratio propter quid*) and its role in sixteenth-century astronomy, see Barker–Goldstein (1998, 243 ff.).

98 Maestlin (1619, 20). The illustration is similar to the one in the treatise on the comet of 1577 showing the parallax of the comet. See Maestlin (1578, 15).

> the stars. We must also deal with comets in the same way if we wish to be certain, whether they dwell in the elementary or the ethereal region, down here below the Moon among the elements or up above in the heavens. For if they are in the elementary region, so must it surely follow that they are subject to a greater degree of parallax than the Moon. If they are up above, however, their measure of parallax could not be as great as that of the Moon. Should we discern in them no difference in parallax or entirely none at all, then it must be the case that comets are so far away that half the Earth's thickness bears no sensible proportion to their distance.[99]

Chapter 6 applies this mathematical procedure, though with a less mathematical focus than in the treatise on the comet of 1577 and with a marked pedagogical approach, to the location and motion of the comet according to the observations of 12 and 13 December. Maestlin concludes that the comet exhibited no observable parallax and, as a consequence, was even more distant than the Sun: "We may conclude irrefutably from the above analysis that the comet was so far removed from the Earth in the ethereal region that any hint of parallax was lost entirely, and therefore that it was situated far above and beyond the Earth than the Sun is from the Earth".[100]

Chapter 7 (pp. 26–35) elaborates upon the above result at length, adding that every comet that had appeared between 1577 and 1607 was located above in the heavens.[101] Maestlin affirms this solemnly from his own observations, but

99 Maestlin (1619, 22). Measurement of the parallax was also praised by Habrecht as the surest way of determining the place or position of the comet. The comet of 1618, he insisted on several occasions, showed practically no parallax. See Habrecht (1618, 13): "zu einem unuberwindlichen beweiß ist die Parallaxis"; 46 f.: "Dieser Comet ist so wol wegen seiner geringenn parallaxi […] nicht weit under dem Marte gestanden"; 52: "Da war kein andere entscheidung, kein Richter unnd Außtrager der sachen als die Mathematischen Instrumenten dardurch man fleißig suchen müssen, ob er auch ein Parallaxin habe". The same praise of the measurement of parallax can be found in Schickard's treatise. Cf. Schickard (1619, 56 ff.) with several images, and 160–162 for the demonstration that the comet of 1618 "schier keine, oder doch gar ein kleine und geringe *Parallaxin* müsse gehabt haben" (160).

100 Maestlin (1619, 24–25): "Aus disem allem ist unwidersprechlich zu schließen, das diser Comet von der Erden *in aethereament regionem* so hoch erhebt geweßt sey, da sich alle *parallaxeon sensibilitas* verliert, und deswegen von und ob der Erden ferner und höher gestanden, als die Sonn von der Erden ist".

101 *Ibidem*, 27: "it is certain that since the year 1572 all comets and new stars, namely the comets of 1577, 1580, 1582, 1585, 1590, 1596, and 1607, along with the stars of 1572 and 1604, have been in the ether, and by no means in the elementary region. For there was no sign of parallax in any of these".

declares that it was probably also the case with the comet of 1556 as well as with those of the 1530s. In contrast, he condemns the attempts of other astronomers to locate comets in the sublunary region (such as Johannes Vögelin with the comet of 1532, and Andreas Nolthius, Bartholomaeus Scultetus and Thaddaeus Hagecius with the comet of 1577) as self-contradictory and plainly wrong.[102]

All this allows Maestlin to state universally—referring to a future, never written cometary treatise—that all comets are celestial appearances.[103] Moreover, the very same comet Aristotle mentions as observed in his own time[104] was, contrary to what Aristotle says, located in the heavens, as its motion seemed to indicate.[105] This result was even more significant for Maestlin, since the comet had apparently led Aristotle to conclude wrongly about the sublunarity of all comets.

How, then, Maestlin asks in Chapter 8, had Aristotle made such a mistake, without present-day Aristotelians ever acknowledging it? Here, Maestlin responds to Keckermann's conclusion that celestial comets mark the ruin of natural science and philosophy as a whole and seeks to save Aristotle by minimising the effects of the cosmological revolution. Aristotle must be excused ("ist wol zu entschuldigen"),[106] Maestlin argues, because he found the older doctrine of celestial comets to be insufficiently grounded, and though his own doctrine was equally weak, he proposed it undogmatically, at best as a 'probable conjecture' before he could propose a better one ("er dise meinung nit lenger begere zu bestreitten, als bis er eins beßern berichtet werde").[107] This is confirmed by Aristotle's declaration right from the beginning:

> To be sure, he admits that he did not hold his opinion completely for certain or firm, and was not fully confident in it, for he complains about it in Book 1, Chap. 7 of the *Meteorologica* and expressly objects to it as well. "We therefore consider," he says, "what one wishes to say about those things

---

102 *Ibidem*, 27–29. For Vögelin, see Kokott (1994, 135 f.); for the others, see Hellman (1944, 216–225 [Nolthius], 206–216 [Scultetus], 184–206 [Hagecius]). For Scultetus, see also Jerratsch (2020, 243–253).

103 *Ibidem*, 30: "Aber hievon soll in eim besondern Tractat, ob Gott woll, ausführlich gehandlet, und angezaigt werden, das vil mehr zu schließen were, **Es seye kein Comet iemals in *regione elementari*, sonder alle samptlich in *aetherea regione* erschinen**".

104 *Meteorologica*, I, 6, 343b 3–5.

105 Maestlin (1619, 30–32).

106 *Ibidem*, 36.

107 *Ibidem*, 38.

that lie hidden from our senses as explained wisely or in a sufficiently reasonable way, should one progress in this endeavor as far as possible".[108]

Aristotle could not, however, apply his empirical appeal to establishing the fact (*oti, quia*) of the actual distances of comets and later attend to the causes (*dioti, propter quid*), because the only way to do this with complete certainty—the doctrine of parallax—was unknown in his time: "That Aristotle was not familiar in his own day with the doctrine of parallax, which could have cleared up the entire matter, shall now be shown. 1. In none of his books does Aristotle make reference to it with a single word nor express a single hint about it".[109] Contemporary Aristotelians, on the other hand, had no excuse for their error, given the evidence accumulated over the last ninety years demonstrating that all comets are celestial bodies.[110]

According to Maestlin, Aristotle was also led astray by the erroneous presupposition of the immutability of the heavens. This had wrongly suggested that comets necessarily belonged to the elementary region. Accordingly, comets

108 *Ibidem*, 36: "Das er aber dise sein mainung nit so gar für gewiß und stählin halte, sonder deren selbs nit gnugsam trawe, zaigt er an, da er sich *lib.* 1. *cap.* 7 *Meteor.* deßen beklagt, und austrucklich gnug protestiert. "Wir haltens es darfür", spricht er, "was man von den jenigen dingen, die vor unsern Sinnen verborgen sind, reden will, so seye es weißlich oder verständtlich gnug erklärt, wann mans so weit, als es müglich est, bringet". Cf. Aristotle (1984), *Meteorology*, I, 7, 344a 5–8. The hypothetical, conjectural character of meteorology was widely acknowledged within the Aristotelian tradition, both in the Middle Ages and in the Renaissance, relying frequently on the same passage from Aristotle's *Meteorologica* I, 7, invoked here by Maestlin. On this, see especially Martin (2011, Chapter 1: "The Epistemology of Meteorology").

109 Maestlin (1619, 38): "Das aber *Doctrina Parallaxeon*, welche dises alles hette könden richtig machen, zu zeit des Aristoteles noch unbekandt gweßt seye, ist daraus zu beweisen. 1) Aristoteles in allen seinen Büchern thut deren mit keinem eintzigen wort meldung, thut auch kein Andeuttung darauff". Schickard speaks in similar terms. See Schickard (1619, 58–59): "Dise *parallaxis* ist ein solche edle künst und herrlich kleinot, das wirs umb vil gelt nit manglen solten. Das seind die rechte *Alae* flügel, damit wir uns in hohen himmel hinauff schwingen [...], die rechte *scalae*, laittern daran wir hinauff steigen müssen; die den alten *philosophis* vor zeitten noch unbewust gwesen, und gemangelt haben, darumb sie von der Cometen höhe so ungereümt geschriben: hatt sampt andern geheimen künsten zu unserer letsten zeitt sollen versparet werden und iez erst geoffenbart. [...] es sey die messung der Cometen höhe nit allein müglich, und menschlicher vernunft zugelassen, sonder auch, wan mans anderst recht fleissig practicirt, gwiß und ohnfehlbar, und also den *Astronomis*, welche die Cometen uber den Mond hinauff setzen, vil besser zu glauben, als den *physicis*, die darvon nichzit wissen könden". For the Melanchthonian origin of the image of arithmetic and geometry as the "wings of the soul", see Melanchthon (1999, 93); Granada (2019, 209), with further references.

110 Maestlin (1619, 39).

were formed from terrestrial exhalations elevated to, and set aflame, in the uppermost region of the air. Aristotle's cometary theory was driven by the premise of the immutability of the heavens, which was not the result, but the principle, reducing comets to sublunary phenomena. As Maestlin clearly affirms:

> The answer shall thus be (though from a presupposition that one ought to accept for certain even if it has not yet been proven): since comets could not exist in the heavens, for there would be no change there, they must abide down here, below the sphere of the Moon among the elements, particularly in the upper air since none of them ever approaches us on earth. Now, they say, since we know the dwelling place of comets, it may be easy to conjecture and estimate where they have their origin, as every form of change is found in elementary nature: and it is evident that many exhalations arise every day from the earth and water, with some turning into water and rain, while others yield fire, lightning, thunder, rays and so forth, such that it is worth considering that many of the exhalations ascend even higher and assemble together in the upper air, like a mountain, until they are parched and prepared, as it were, to be ignited by the element of fire, which lies just above, and by the swift | circuit of the heavens; and since the heavens move around with them by force, as it were, the sphere of fire, together with the upper part of the air, comets must also move around with them, and so forth.[111]

Contrary to Keckermann, Maestlin further affirms that natural science and philosophy are not undermined by the celestial location of comets. Rather, wrong preconceptions are replaced by true explanations on the basis of reliable experience:

> Since there are now so many new things that appear every day about which Aristotle and others knew nothing or even dreamed, should Aristotle's philosophy therefore be all the less appreciated or thrown away entirely? Not at all, rather one should hold it with due respect and give thanks to God for it. Whatever else is added to it one should certainly not reject, either, but only praise all the more God's inexpressible wisdom, excellence, and omnipotence. That is why Keckermann and others like him should have been kept from airing their unnecessary complaint and

111 *Ibidem*, 39–40.

> rather recalled that even in our age, as Seneca says in the *Naturales quaestiones*, Book 7, many things remain hidden that our descendants shall discover after we are forgotten.[112]

Reliance on experience for a better causal understanding of meteorological phenomena, an epistemological position agreed upon by Aristotelian natural philosophers (as clearly demonstrated by Craig Martin),[113] was now extended by Maestlin to an issue that Aristotle and his philosophical followers were not ready to abandon: the immutability of the heavens. Meteors and, as a consequence, comets were the subject of deep discussion among Aristotelian philosophers during this period. This was not the case, however, with the celestial location of comets and the mutability of the heavens, since the latter point was a fundamental tenet of Aristotelian cosmology. Thus, for Aristotelians celestial immutability necessarily entailed, except for phenomena resulting from supranatural divine intervention, as Keckermann conceded,[114] the sublunary character of comets. Comets were thus thought to be ephemeral bodies lacking the regular and perfectly ordered motion of the heavens.[115] Maestlin

112 *Ibidem*, 52 f.: "Weil nun so vil Newes täglich herfür kompt, darvon Aristoteles und andere nichts gewüßt, inen auch darvon nie getraumet hatt, solte darumb *Aristotelis Philosophia* desto geringer zu schätzen, oder gar hinzuwerffen sein? Mit nichten, sonder man soll sie in billichen Ehren halten, und Gott darfür dancken. Was aber weitter hinzukompt, soll man freylich auch nit ausschlagen, sonder Gottes unaussprechliche Weisheit, Gutte, und Allmacht nur desto mehr preisen. Deswegen hette Keckermann, und andere seins gleichen, der unnotwendigen klag wol sollen überhebt sein, und vil mehr gedencken, das noch zu unsern Zeitten, wie Seneca *lib.* 7 *quaest. Nat.* sagt, vil sachen verborgen seind, welche unsere Nachkomme, wan man unser nicht mehr gedencken wirt, erfinden werden". For the corresponding passage in Seneca see Seneca (1972, VII, 30, 5), quoted also by Maestlin below, p. 45, note 130 in this Introduction.

113 Cf. Martin (2011, 35 f., 64 ff.).

114 Cf. Keckermann, *Diaskepsis*, in Keckermann (1606, 399–400): "Neque enim consequens est: Apparuerunt in caelo novae stellae; *Ergo*, doctrina, quae ab Aristotele, et aliis Physicis tradita est de cometis et eorum generatione, itemque de caelestis corporis puritate et perpetuitate, veritatem suam amplius tueri nequit. Possunt namque in caelo nova corpora produci per miraculosam Dei potentiam, ut tamen nihil interea ordinariis naturae principiis, viribus, et agendi modis decedat. [...] Nusquam sane negavit Aristoteles, primam causam posse agere sine secundis, et aliter quam secundas; nusquam inficiatus est, Deum posse aliquid novi producere; nec sequitur, *Deus saepe absque natura agit extraordinarie, ergo cum natura non agit ordinarie*; Nec denique sequitur, *Deus saepe facit miracula, ergo evertuntur Physicae scientiae principia et praecepta de natura caeli, deque generatione Meteororum*. Sunt supra naturam miracula, nec tamen naturam oppugnant aut evertunt" (translated above, p. 29).

115 The proportional motion of the comet of 1618, as well as of comets in general, was strongly emphasised by Habrecht (1618, 12, 45, 47), by Schickard (1619 section 39, 110): "Solche Bewe-

could have verified this not only in the works of Keckermann, but also in many other cometary treatises. Suffice here to mention one treatise by Abraham Rockenbach (1536–1611), published in 1602 and reprinted in 1619, whose first chapter ("De divisione mundi") establishes as an indubitable truth the principle of cosmological dualism.[116] As a consequence, when Rockenbach deals in his third chapter with the "nature of comets", he rejects their celestial location as a recent renovation of the old Pythagorean conception[117] and dismisses it out of hand as incompatible with the extreme mutability of comets: "Can it be possible that any celestial body should undertake such different and varying alterations, as appear in the form of comets, and we seize upon in other accidents?".[118]

8 Sunspots and the Telescope Appear on the Scene

We see, then, that the more physical and philosophical nature of Maestlin's cometary treatise of 1619, apart from the limitations imposed by the addressee, is a consequence of his aim to respond critically to the Aristotelian doctrine which remained prevalent among natural philosophers and in the teaching of the schools.

In order to carve out a space for the new concepts on comets and to abandon celestial immutability as a false notion that excluded comets from the celestial region (except in the form of divine miracles), Maestlin appeals in Chapter 9 to the evidence given by the telescope for the mutability of the heavens and the similarity between the earth and the Moon and even the Sun. Through the telescope, Maestlin argues, we may observe that the surface of the Moon is like

---

gung ist gemeinlich Regularis ordenlich und Richtig", and by Crüger (1619, 75): "So ist nun erstlich von geübten und der sachen verstendigen *Astronomis* befunden/ das der Cometen lauff nicht *irregularis* sondern *circulo maximo perfectè subordinatus* sey/ welchs alleine den himmlischen Cörpern und keinen elementarischen zustehet".

116 See Rockenbach (1602). On its contents and the 1619 reprint, in a joint edition with the treatise by Philip Müller, see Gindhardt (2006, 222–243).

117 Rockenbach (1602, 23): "Multi ab ea [Aristotle's theory of comets] recedentes, ad Pythagoreorum opinionem *inclinant*, ut malint credere, Cometas esse vel unam, vel plures perpetuas stellas" (our italics). Rockenbach does not consider that most contemporary supralunarists, such as Brahe, Kepler and Maestlin, believed that celestial comets were temporary productions. Cf. Gindhart, (2006, 227).

118 Rockenbach (1602, 24): "An verisimile est, ullum posse esse coeleste corpus, quod tam varias, ac dissimiles mutationes recipiat, quales sunt in figura Cometarum, et aliis accidentibus deprehendimus?".

that of the Earth, with mountains, valleys and seas. Thus, it is reasonable to conclude that habitation there, like here on Earth, is probable:

> And since it is not probable that the All-Wise God would have created in vain mountains and valleys also on the Moon—"God and Nature," the philosophers state, "do nothing in vain," which is to say that God and Nature act, perform, and produce nothing for no reason or in vain—one might thus further argue with good reason that the same would apply to any form of change [*alterationes rerum*], and thus to generation and corruption as well, and so forth. If the ancient philosophers could have seen this as clearly as we do now through such optical looking glasses, they would have no doubt been fortified in their views even more firmly, since they believed there are many cities and mountains on the Moon, as Xenophanes teaches, and that it is inhabited all around by rational people and irrational animals, in the same way as we witness here on earth, and adorned with much more beautiful plants (trees and fruits), as the followers of Pythagoras said, for why else should this be without the other? And so forth.[119]

Maestlin suggests that the same (without excluding habitation) can be said for the Sun.[120] Moreover, the recently discovered sunspots—Maestlin refers explicitly to the treatise by Johannes Fabricius (1587–1616),[121] the first printed tract on the argument, making no mention of Galileo or Christoph Scheiner—confirm the mutability of the Sun.[122] In addition, Maestlin adduces several biblical quotations in support of the impurity of the heavens:

> one may thus well consider that these spots and mountains on the Sun and Moon were not created in vain, but that among and along with them

119 Maestlin (1619, 51). Cf. also p. 58.

120 *Ibidem*, 58.

121 Fabricius (1611). Cf. Maestlin (1619, 52, 55). For a recent study on the earliest discoveries and discussions of sunspots, see Galilei–Scheiner (2010, 30–34 on Fabricius).

122 Philip Müller also refers to sunspots as evidence against Aristotle. However, he proposes, echoing Kepler, that they serve as a material source for the "new celestial phenomena" instead of the Milky Way. Cf. Müller (1619, 49): "Et quia tubis dioptricis cotidie deprehenditur aliquid nigrarum rerum, quod obvolitet soli et ceu commaculet atris punctis splendorem ejus, idcirco perspicacioribus ex Keplero cogitandum propono, an non inde commodius, quam ex via lactea, materia gignendis novis in coelo corporibus depromi queat". On Kepler's role in the dissemination of this hypothesis, also assumed in the Netherlands by Willebrord Snel, see now Mehl (2013, 242 ff.) and Mehl (2019, 318–328). For Snel's views, see Van Nouhuys (1998, 354 ff.).

still more things were created, for we should certainly not believe that God unleashed his endless power only and utterly on this single body, the Earth, where we reside, but what He has done here on the Earth (which is only a small point with respect to the upper heaven), should nothing further have remained for him, and could not every deed and wonder (although they are without number, Job 5:9 and 9:10, and one can scarcely discern a small fraction of them, Sir. 42:23), which he performed here not also have been performed there on the Sun and Moon, or even greater things as well?[123]

Maestlin now introduces a digression on sunspots and considers the following three points in succession:

1) Measuring parallax allows us to conclude that the spots are not located in the sublunary air, but far beyond and, in fact, on the Sun.
2) Though these phenomena have been observed only in our time thanks to the telescope, they are by no means new phenomena that stem from the *senectus mundi*; rather, they are permanent, natural phenomena on the Sun; and the presumed observation, shortly before Charlemagne's death, of a solar transit of Mercury was doubtless a sunspot.[124]

---

123 Maestlin (1619, 58): "so mag man wol erachten, es seyen dise Mackeln, und Berg an Sonn und Mon nicht vergebens, sonder neben und mit inen, auch noch mehr anders geschaffen, dann es ist ja nit zu glauben, das Gott sein Allmacht habe allerdings und volkommen in disem aintzigen *corpore*, in der Erden, da wir wohnen, ausgelehret, sonder was Er alhie an dise Erden (welche gegen dem hohen Himmel nur wie ein Püncklin ist) gewendet, solt er ime nichts weitters vorbehalten haben, und alle Wunder und thatten (wiewol sie nit zu zehlen sind, Iob 5. v. 9. *et cap.* 9. v. 10. und man kaum ein püncklein davon erkennen kan. Syr. 42. v. 23.) was er hie geschaffen, nicht auch dorten an Sonn und Mon, schaffen könden, oder noch größers zu schaffen vermögen?".

124 Maestlin (1619, 54–56). Maestlin also refers to his discussion of this presumed transit of Mercury in his *Disputatio de multivariis motuum planetarum in coelo apparentibus irregularitatibus*; see Maestlin (1606, 54–58). The issue had been a source of conflict with Kepler. As is known, Kepler presented in his *Phaenomenon singulare seu Mercurius in Sole* (Leipzig, Schurer, 1609; *JKGW*, iv, 80–98) his presumed observation (on 18/28 May 1607) of Mercury's solar transit, which confirmed an earlier observation of it in 807 by an astronomer at the court of Charlemagne, as reported by Kepler in his *Optics* (1604, 306; *JKGW*, ii, pp. 264f.). Kepler's reference to the observation of Mercury's transit in 808 had been rejected, without mentioning him by name, by Michael Maestlin and Samuel Hafenreffer in the abovementioned *Disputatio*, thesis xcviii, pp. 54–58. After receiving a copy of it along with a letter by Hafenreffer (cf. *JKGW*, xv, nº 396), Kepler answered with a letter of 16 November 1606 (cf. *JKGW*, xv, nº 400) in which, besides defending his new celestial physics, he restated his opinion: "De Mercurio sub Sole non persuades", line 28. In *Phenomenon singulare*, Kepler once again reaffirmed his position. However, he later recanted

3) The question whether the spots are moving around the Sun, as clouds do around the Earth, or on the solar surface is insoluble, even if Maestlin inclines towards the former possibility:

> we shall ask here what these black masses or spots in the Sun are, and whether they are found on the Sun itself, like the spots on the Moon, or whether they are separate from the Sun, floating around it like the clouds around the Earth. Answer. This, I believe, shall surpass the height of human reason, for the Sun lies too far away from us to fathom it. However, should this and other such matters still be submitted for our evaluation, since it is said, "The heavens are not pure for God," and "Behold, even the Moon is not bright and the stars are not pure in his eyes", I leave it for more insightful minds to consider. On the other hand, as long as human curiosity is not inopportune, but employed to praise the endless power and wisdom of God, it is not forbidden from reflecting on these secrets a little further. For example, since not only some of these spots appear larger than the largest stars, but also the mountains on the Sun [sic] are so apparent that they are seen and noted here below on Earth at such an extremely great distance, it may thus be gathered how great they are in their own right. Next, since these spots appear in so many different forms, now one, then two, three, or more in greater number, some small, then large, some round, some assembled together and then separated and scattered, we may reckon from this that they are not in or on the Sun but travel around it, for otherwise they would remain the same, just like the spots on the Moon; however, they do have an analogy and comparison with the clouds, except that they remain on the same route, while the clouds travel back and forth about the Earth without any sense of order, according to elementary nature. This and other such matters we may conjecture and consider.[125]

Besides expressing some skepticism about man's capacity for solving these questions, the above passage also allows us to appreciate that Maestlin maintains, beyond their acknowledged mutability, that the heavens are superior and pure when compared with the sublunary realm.

(in his *Ephemerides*, printed in 1618; see *KGW*, xi/1, pp. 28–30), under the assumption that what he had most probably seen in 1608 was simply a spot on the Sun before the advent of the telescope.

125 Maestlin (1619, 56–57).

Consequently, Maestlin refuted celestial immutability and the substantial difference between the heavens and elementary world by experience. In addition, the celestial location of comets rendered implausible the claim that they might originate from terrestrial exhalations elevated to the heavens and there set on fire. Maestlin further claimed that the great magnitude of the recent comet, had it originated from earthly exhalations, would never have allowed for the undiminished permanence of our planet. As a result, the matter of comets could not be terrestrial:

> how or whether the "*exhalationes vel fumi*," the vapors, exhalations, or fumes, that is to say, that matter from which comets are generated, could rise from the earth so high up in the heavens, even higher than the Sun. In answering this question, we must bear in mind what we now know for certain, that comets appear and abide in the upper heaven, in the ethereal region, so that it is not only incredible, but (speaking naturally)[126] absolutely impossible that the exhalations or vapors arising from the Earth should be able to ascend all the way there in such an incomprehensible heap without any perceivable loss of land. For, first of all, consider the immeasurable magnitude of the present comet, which, as said above, was more than 1,000,000 miles long and over 100,000 miles wide; how great, then, was its body in its own right? Certainly, the entire globe of the Earth, as big as it always is, if it were resolved entirely into exhalations and vapors, should scarcely be large enough for such a seemingly great body [the comet].[127]

---

126 In the sense of the scholastic formula "loquens ut naturalis", that is, speaking according to the philosophers, regardless of revelation and the faith. See Bianchi–Randi (1990, 33–56).

127 Maestlin (1619, 58–59). Cf. also 61–62: "Aus disem ist jetzt abzunemmen, wie vil tausent und aber tausentmal größer, als die Erde, diser unser Comet muß geweßt sein? Wie wolte es dann müglich sein, das aus und von der Erden köndte so gar vil Dampf oder Rauchwerck, zu einen solchen auf ettlich Monat weerenden Cometen gnugsam ausfahren, ohne wol empfündtlichen deren Abgang? Es müßte ja nicht nur ein Berg oder zween, nicht nur gantze Länder verzeeret werden, sonder so auch der gantze Erdbod würde verstieben und verriechen, möchte es doch kaum gnug sein, oder wans schon dis were, würde doch von der Erden nit vil mehr übrig sein. Nun aber bleibt der gantze Erdkraiß mit Waßer und Landen in vorigem Stand und Wesen, und wirt in aller Welt kein münderung gespüret. Darumb ist zu schließen, das die Cometen ire Uhrsprüng, substantz und Materi mit nichten aus und von den Düfften, so von Erden auffsteigen, bekommen, sonder anderswoher haben". The Aristotelian doctrine of the origin of comets from terrestrial exhalations inflamed in the upper air was also firmly rejected by Habrecht (1618, 9, 13, 28, 50) and by Schickard (1619, sections 2, 4, 21–25).

Maestlin considered it equally impossible that comets could be inflammations. In all probability, he pondered, no sphere of fire ever existed. In any case, whether they were incensed or illuminated, terrestrial exhalations destined to become comets should necessarily become visible long before their final inflammation or illumination, when they rose to the heavens:

> given, though not granted in the least, that they could arise from the Earth and assemble in such an immeasurable mass without any damage, diminution, or loss to it, the question now is whether they ever reached that high up. Here, we must answer on the basis of optics simply with **No**. The reason for this answer is that they would have been seen up there even before they were ignited or had begun to shine brightly. This has never happened, however, or we would have heard of such a thing.[128]

Maestlin tries to demonstrate the above claim by recalling Alhazen's *De crepusculis* and Witelo's *Optica*, but the manuscript breaks off before his argument concludes. Here, Maestlin's attitude contrasts markedly with contemporary authors such as Kepler or Snel, who dared to concede to planets and even the Sun the emission of exhalations that naturally gave birth to celestial comets.[129] Maestlin does not provide any positive statement concerning the matter or generation of comets. Just as he had refused to accept Kepler's celestial physics, arguing for a purely mathematical description of celestial motion, Maestlin remained rather skeptical about man's ability to arrive at any certainty when it came to cometary physics. Along with so many of his contemporaries he preferred to invoke the well-known words of Seneca leaving a more advanced

---

128 Maestlin (1619, 62): "So ist ietzt die frag: Ob sie jemals so hoch hinauf kommen seyen? Hierauf ist ex *Opticis simpliciter* mit **Nein** zu antworten. Ursach dieser Antwort ist, dann sie würden auch vor irer entzündung, oder ehe sie zu leuchten anfiengen, droben gesehen werden. Solches ist aber niemals geschehen, oder das es iemals geschehen were erhört worden".

129 On Kepler, suffice here to refer the reader to Boner (2013, 127 ff.); on Willebrord Snel, see Van Nouhuys (1998, 337–360). See also Mehl (2013, 246 f.) and Hamel (2016) on Simon Marius and his debt to Kepler. By contrast, this is a point never mentioned by Habrecht, who preferred to consider the Milky Way as the source of cometary matter; cf. Habrecht (1618, 14–15). Quite to the contrary, Schickard excluded the Milky Way as the source of cometary matter, claiming that comets were born and moved among the planets, far away from the stellar sphere; see *Cometen Beschreibung*, section 3 ("So werden die Cometen auch nit auß der Galaxia oder Milchstraß des Himmels geborn; dan dieselbige vil mahls höher ist, als die Cometen steigen"), pp. 11–13. According to Schickard, "die Materi der Cometen, ist Aetherisch, gleicher Substantz mit dem himmel, doch ettwas dickers und besser gedigen, und dannoch durchsichtig", *ibidem*, p. 13.

knowledge to future generations[130] or simply admitting the limits of human knowledge: "This and other such matters we may conjecture and consider, but what Seneca says about comets remains: 'There are many other things that pass by in secret and do not appear before our eyes. God has not revealed all things to the eyes of men'".[131]

In any case, Maestlin's change of strategy from his treatises of 1578 and 1580, even when considering the differences in their addressees, bears witness to the broad resistance of the academic establishment to the new cosmological ideas necessarily connected to the explanation of comets as purely natural phenomena. Last but not least, this resistance expressed the desperate attempt of natural philosophers in the schools to maintain the distribution of competencies among the established disciplines and refuse mathematicians any right to determine the essential properties of natural phenomena.[132]

9 Maestlin, Schickard and Habrecht on Faulhaber and the Rosicrucians

In 1617, the Ulm mathematician Johannes Faulhaber published a calendar (an *Einblattdruck-Kalender*, printed on a single folio), in which he prophesied the appearance of a comet the following 1 September 1618 (O.S.).[133] This prophecy

130 Seneca (1972, VII, 30, 5), cited on p. 38 above: "Multa venientis aevi populus ignota nobis sciet; multa saeculis tunc futuris cum memoria nostri exoleverit reservantur" ["Many things that are unknown to us the people of a coming age will know. Many discoveries are reserved for ages still to come, when memory of us will have been effaced"]. Seneca's words were also quoted by Kepler in the frontispiece of his *De cometis libelli tres* ("Erit qui demonstret aliquando, in quibus Cometae partibus errent, cur tam seducti a caeteris eant, quanti qualesque sint. Contenti simus inventis: aliquid veritati et posteri conferant", *JKGW*, viii, p. 132) and by Schickard in his *Kometen Beschreibung*, p. 111. Schickard, who often quotes the Roman author approvingly, calls him "mein Seneca", p. 70.

131 Maestlin (1619, 57): "Dises und dergleichen [about sunspots] mag man coniecturieren und schätzen, doch bleibts darbey, was Seneca von Cometen sagt: 'Es seind vil sachen, welche im verborgen fürübergehen, und kommen menschlichen Augen nit ins gesicht. Gott hatt der Menschen Augen nit alles geoffenbaret'"; the quotation is underlined in the original text. We have modified the English translation by reading, with Maestlin, *patefecit*, as implied by German *geoffenbaret*, instead of *fecit*.

132 On this subject, see Granada (2019).

133 On Faulhaber and the issue of the comet, see Schneider (1993, 15–22); Hawlitschek (1995, 44–50); Schneider (2004); Gindhardt (2006, 87–102). On the possibility of a personal meet-

or prognostication was founded upon Faulhaber's mystic-cabalistic number speculation, whose mathematical foundations Faulhaber had published in several studies produced in previous years.[134] When he perceived the presence in the heavens of the first comet of 1618, in August, Faulhaber boasted that his prophecy had come true. In addition, Faulhaber connected his conception of the mystical numbers and the prophecy of the new comet with his eschatological and chiliastic hopes for the near future, as well as with the recent Rosicrucian manifestos, printed in Kassel in 1614 and 1615. Faulhaber had, in fact, sought to contact the secret fraternity that had published the manifestos announcing the imminent onset of a new era in the history of humanity. The prophesied comet and the eschatological claims came directly to the fore in Faulhaber's *Fama syderea nova*, printed for the first time in Nuremberg in December 1618, while the third comet was still visible.[135] In this work, whose title recalls Galileo's *Sidereus nuncius* as well as the Rosicrucian manifesto *Fama fraternitatis*, Faulhaber linked the comet of 1618 to the Bohemian revolt against Habsburg rule. In this way, his prophecy adhered to the interpretation of the new Bohemian King Friedrich V of the Palatinate (1596–1632) as the lion destined to overthrow the Habsburg eagle according to the prophecy of 2 Esdras (4 Ezra).[136]

On 26 August 1618, Faulhaber asked his friend Matthäus Beger, a mathematician in Reutlingen, to request from Maestlin his observations of the first comet that appeared that year.[137] Maestlin answered only on 18 January

---

ing between Faulhaber and Descartes in 1619, see also Gäbe (1972, 12–19); Hawlitschek (1995, 55–81) and Mehl (2019, 211–240).

134 Faulhaber (1604), Faulhaber (1612), Faulhaber (1613), Faulhaber (1613b), Faulhaber (1613c). See Schneider (1993, 229 ff.). Faulhaber's critics later related this presumptive prophecy to the indication in Kepler's Ephemerides for 1618 (J. Kepler, *Ephemerides Novae Motuum Coelestium ab anno vulgaris aerae MDCXVII*, Linz, Johann Blanck, s. d.; *JKGW*, XI/1) of the identical longitude of Mars (3° 33′) with the latitude of the Moon on 1 September 1618 (old style). In his *Prognosticum Astrologicum auff das Jahr MDCXVIII* (Linz, Johann Blanck, 1618; *JKGW*, XI/2, p. 170. 5–6), Kepler had also posited the probable appearance of a comet, since the last one had appeared in 1607.

135 Faulhaber (1618–1619). The editor was Daniel Mögling (1596–1635, under the pseudonym Julius Gerhardinus Goldtbeeg Jenensis), an admirer of both Faulhaber and the Rosicrucian fraternity. In the same year, Mögling had published the famous *Speculum Sophicum Rhodo-Stauroticum* in Frankfurt.

136 On this subject, see Gilly (2002). Cf. Mehl (2013, 236 ff. and 240) and Mehl (2019, 315 f.) for the famous engraving illustrating all these motifs. For the apocryphal book of 4 Ezra, see Hamilton (1999) and Granada (2017).

137 As acknowledged by Beger in his letter of 27 January 1619 that opens Beger (1619). This letter was also printed at least with some of the editions of Faulhaber's *Fama*, just as that printed in Nuremberg quoted above (Beger's letter in sig. Di^v and the *Problema* in

1619, just a few days after receiving from Beger Faulhaber's calendar containing his famous prophecy, as well as another writing by Beger (simply a letter or, perhaps more probably, the *Problema Astronomicum*). In his response, recently published by Édouard Mehl,[138] Maestlin refuses to consider the astrological significance of the comet, adducing, as he always did, that he had never actually practised the art.[139] However, Maestlin agrees about the dreadful impact of recent events in Bohemia.[140] Without reservation and in an ironic fashion, he unleashes his full scepticism concerning the cometary prophecy:

> Nun sihe Ich in des Herrn Calender (für dessen überschickung Ich mich freundtlich bedancke [...]) das der Herr nicht nur Andeuttung des Cometen gethon, sonder laße mir sein dexterität, das er selbigen errathen vnd getroffen, wiewol kein fundament, daraus Er, das ein Comet erfolgen möchte, judiciert, gesetzt wirt, wolgefallen.[141]

---

sig. Dii^r–iv^v). Beger added in his letter that he had also sent to Maestlin a copy of Faulhaber's calendar for 1618 containing the prophecy of the comet that had appeared in the meantime. The *Problema* consisted mainly in a German translation of the description of Maestlin's method for observing the nova in Cassiopeia, as described in Maestlin's treatise published in 1573. See Maestlin (1573, 30–32). On Beger, see Hawlitschek (1995, 216–222); Betsch (1996, 145–148), Hawlitschek (2006, 109–113).

138 Mehl (2013, 254 f.). A previous edition of this letter was provided by Hawlitschek (1995, 304 f.).

139 Mehl (2013, 254): "Ich kein Astrologus bin, mich auch niemals darauf begeben"; 255: "Ich bin, wie oben gemeldt, kein Astrologus, darumb Ich hiervon nichts zu prognosticieren wais".

140 Mehl (2013, 255): "Aber wer die laidige, betrübte zuständ schier in aller welt, vnd sonderlich in Böhem, ansihet, der mag leichtlich erachten, daß ein schweeres Vnglück ob vns schwebe, vnd zu besorgen, es werden die funckhen des, in Böhem angezundten fewrs, weit, weit, vnd gar weit, vmbher stieben vnd fliegen. Gott sey vns gnädig, vnd erhalte vns. Amen". The situation in Bohemia was also of deep concern for Isaac Habrecht and Wilhelm Schickard in the prognostics closing their treatises on the comet; see Habrecht (1618, 57 ff.); Schickard (1619, 168–169).

141 Mehl (2013, 254): "Now I see in this gentleman's Calendar (for whose transmission I warmly give thanks [...]) that he not only foretold the comet, but I delight in his dexterity at guessing and hitting upon it, though no foundation is given from which he judges that a comet might occur" (our translation). Kepler also rejected in his *Prognosticum auff das Jahr 1620* the alleged prophecy of the three comets by the Rosicrucians (whom he claims not to know). See Kepler (1619, 211 f.), transcribed and translated in Mehl (2019, 408–412). According to Kepler, the Rosicrucians would not have seen the first comet and would have mistaken the second and third comets for a 'halo' around Jupiter, which was then at its maximum apparent size.

Concerning astronomical and cosmological matters, Maestlin affirms that the comet (in this case, the third one) was higher in the heavens than the Sun, as evinced by the absence of parallax.[142] The last time he saw it (on 13 January), Maestlin claimed, the comet was in the vicinity of the Little Bear. Here, Maestlin refers Faulhaber to his following publication: "The comet was last visible, though just barely, not far from the Little Bear as, God willing, shall soon be specifically shown."[143]

There is no doubt that Maestlin is referring here to his manuscript treatise on the comet of 1618, which he had still not finished. Significantly, it is in this work, namely the original rough draft, where on page 9 (apparently in the draft of the second chapter) we find a section (see Figure 3) entitled in the margin, "Can comets be predicted?" (*An Cometae prognosci possent?*). Though no name is given, we can safely assume that Maestlin had Faulhaber and his famous prophecy in mind. Here is the passage:

> Here arise various questions about all comets, as well as the present comet in particular, one of which is whether one could prognosticate from the astrological art and science [*ex arte et scientia astrologica*] the appearances of an extraordinary *meteorum*, especially a comet, and proclaim them before they appear. Answer. I am certainly not an astrologer, nor have I ever wished to resort to it for certain reasons. On the other hand, I find that those who practice astrology prudently and with the proper measure and sense know how to uncover many mysteries and secrets of nature, especially what may or tends to occur in the ethereal region, particularly if they remain within the limits of art and science (which unfortunately is not the case with everyone) and do not stray too far. Yet when it comes to the predictions and proclamations of comets, it is my simple scruple that one can well know those things that are astronomical, as well as the certain course of comets in nature, for we know how to calculate with certainty such things as when a planet appears larger at

142 Mehl (2013, 255): "In welcher sphæra cœlesti Er stehe, kan Ich nit judicieren, ohn das es gewiß, das Er nicht in regione Elementari, im Lufft, oder sphæra ignis, vnder des Mons Himmel wie die philosophi ex doctrina Aristotelis halten, gestanden ist. Dann Er hatt kein parallaxin ghabt, wie solches aus den obseruationibus klärlich zu erweisen ist. Deswegen Er one Zweifel höher als die Sonn, seinen lauf ghabt, wie hoch aber droben, kan Ich nit erkennen".

143 Mehl (2013, 254): "Er [the Comet] nit fern von dem kleinen wagen, sich letztes mal hatt, doch kaum, sehen lassen, *wie, ob Gott will, solle in kürtze, specificè angezaigt werden*" (our italics).

> times in the heavens than at others, which has often happened in the case of Mars, when it grows significantly in size, that is, when it reaches perigee on its epicycle and eccentric, and is generally held by the common people to be a new star and even I am called upon, among others, for this reason. Thus, every time Venus moves more than 30° in front of the Sun it can be seen quite well for many hours during the day, among other astronomical matters of this sort. Such things on earth and in the water and air give the aspects of the stars certain radiations [*certas irradiationes*], by which lower elementary nature is directed and ordered. From this, a shrewd and careful astrologer can predict changes in the air and much more. Yet I strongly doubt whether what may not be achieved in celestial matters according to astronomical calculation from the motions may be judged astrologically with any degree of certainty or probability.[144]

Unfortunately, Maestlin never returned to this matter in the sections preserved from the clean copy. As a result, it is impossible to say whether the lost first section (pages 1–8, covering the first two chapters of the treatise) turned back to the topic, although it seems rather improbable.[145]

By contrast, Schickard explicitly addressed the issue in a section of his *Cometen Beschreibung*, rejecting the notion that the appearance of a comet could ever be predicted. This brief section, titled "kein Comet gwiss prophezeit, oder zuvor kan ausser dem gstirn geweissagt werden", with the note in the margin that "Cometas praedici posse negatur", concluded with an openly contemptuous reference to Faulhaber and the Rosicrucians:

> Yet when Hebenstreit further reports on the kabbalistic and numerological prophecies, he no doubt mocks his countryman Faulhaber, who deals quite a bit with this kind of empty speculation. Hebenstreit not unreasonably requests from [Faulhaber] that if he has obtained any such secrets from his cubical numbers and is able to prophesy from them, that he communicate how to judicious people, in accordance with Christian charity, so that the spirits may be tested. In this way, he will win wide renown and great praise. But he should be careful, for if [Faulhaber]

144 For the original German text, see the transcription in Appendix 1 to the present edition.

145 For the ensuing debate on Faulhaber's prophecy and *Fama syderea nova*, see Gindhart (2006, 87–102), and Mehl (2013, 239 ff.).

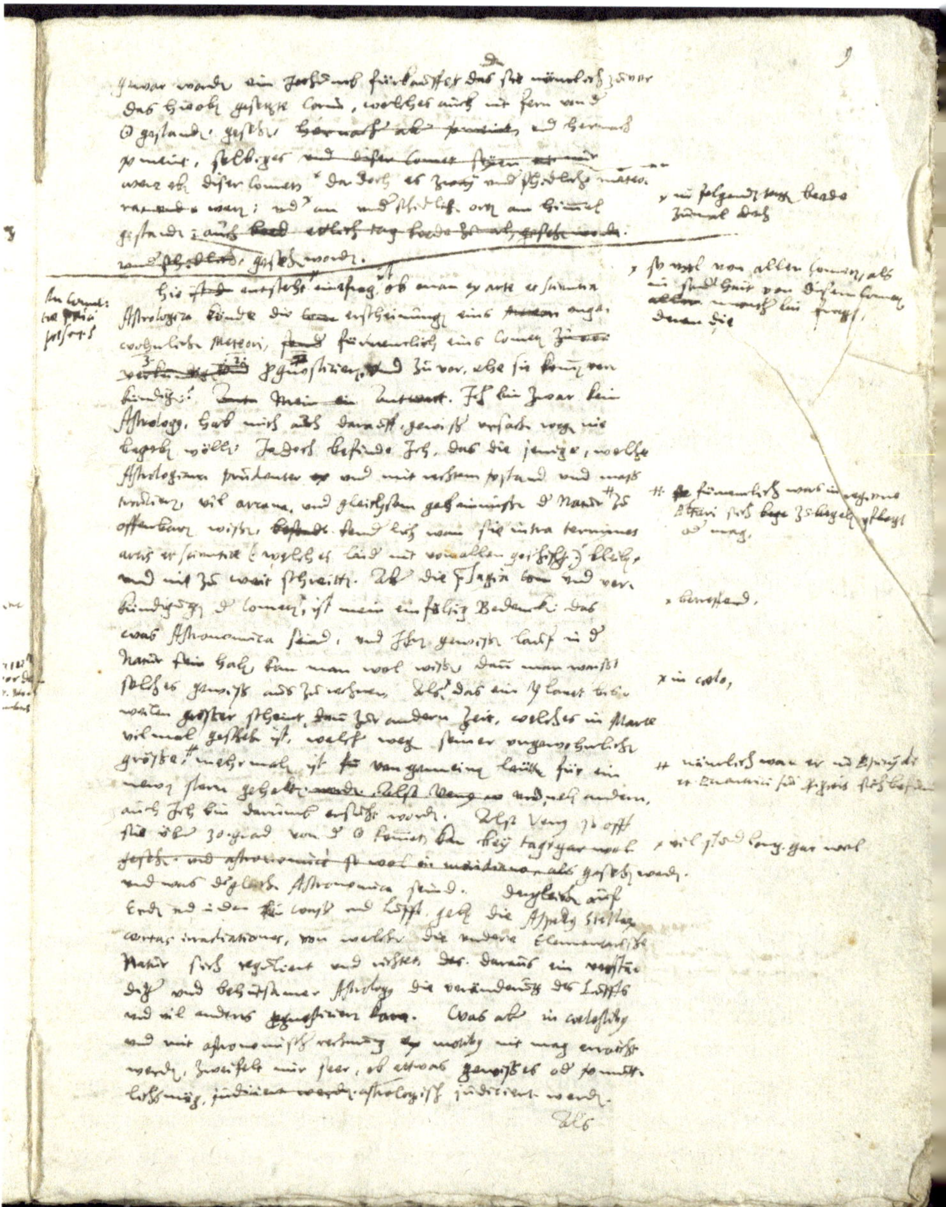

FIGURE 3 Page 9 of the fascicle 8a showing the section on the predictability of comets

> cannot do it, he will not prevent the vanity of his groundless principles from being discovered, and his shame shall come to light. Finally, it is also a wonder that the well assembled Rosicrucian Fraternity does not possess such a prophecy, since it publicly boasts to know everything.[146]

In 1619, Schickard wrote a report on six Rosicrucian writings at the request of the University of Tübingen. His judgment was quite positive, concerning primarily the quest for a reformation of the sciences. Strong criticism was reserved for the mistaken astronomical calculations in their writings.[147] Isaac Habrecht also referred to the Rosicrucians in his *Kurtze und gründliche Beschreibung*,[148] addressing them for an open interpretation of the meaning of the comet in the light of their particular programme:

> This comet then signifies all sorts of unrest and belligerency throughout Europe and a portion of America, although I have no doubt that the Rosicrucian Fraternity will explain this matter in their own way, just as

146 Schickard (1619, 33): "Das aber *Hebenstreitius* weitter auch der Cabalistischen und Arithmomantischen Weissagungen Meldung thut, sticht er ohn allen zweiffel auff seinen landsman den Faulhabern, welcher mit dergleich *jejunis Speculationibus* vil umbgeht: und begehrt nit unbillig von ihm, wan er auß seinen Cubicossischen Zahlen dergleichen gheimnussen herauß zwingen und vorsagen könd, so soll er es, Christlicher lieb gemäß, verstendigen leutten communiciern, das man die geister prüfen mög. So werd ihn meniglich Rühmen und grossen dankh sagen. Sorg aber wol, er köndte es nit thuon, werd es auch nit thuon, das die *Vanitas principiorum* nichtige grund nit entdeckt werd, und mit schand an tag komme. Endlich ist es auch ein Wunder, das nit auch die zusamen rottirte Fraternitet vom Rosenkreutz, solchen Prophezeit haben, weil sie sich ja, alles zu wissen, offentlich vermessen". Schickard is referring to Johann Baptist Hebenstreit (–1638); see Hebenstreit (1618, 14) and Schickard (1619, 167): "auch bey diser verkherten Welt nit leichtlich besser, sonder bsorgich nur erger werden möchte. Derentwegen abermahl *probabiliter* hierauß zu schliessen, es möchten obgesetzte drey Stuckh, *res Ecclesiae, Reipublicae et privatae*, nit in *melius*, sonder *deterius* verendert: Und also die Rosencreutz brüder, und andere Chiliasten, so noch auff ein *aureum seculum* warten, da die Wölff bey den lämmern wohnen, die schwerter zu pfluogscharen gemacht, ihrer hoffnung betrogen werden. Weil solche Weissagungen Esaiae und Michae alberait vor der Zeit aber *longe alio sensu* erfüllet werden". Lucas Osiander Junior (1571–1638), Professor of Theology at Tübingen and avid defender of the Lutheran orthodoxy, also vehemently rejected the Rosicrucian manifestos and Faulhaber's chiliastic hopes founded upon "Arithmeticam nescio quam Cabalam"; see Osiander (1619, sig. A3^{r}).

147 See the edition of the report in van Dülmen (1971, 254–259); we owe the reference for this article to Carlos Gilly. On this report and Schickard's covert sympathies for some aspects of the Rosicrucian manifestos, see Kühlmann (1995, 61); Bubenheimer (1995).

148 Cf. Habrecht (1618, 58, 65–66).

they have recently explained on their own openly in the *speculo sophico*, published this year. May their voice thus be heard [on the matter] however they wish.[149]

149 Habrecht (1618, 65): "Bedeutet also dieser Comet der mehrertheil Unruh und Kriegswesen in gantz Europen und einem theil America: wiewol mir nicht zweiffelt, es werde die fraternitet der Rosen kreutzer solches auff ihr thun außdeuten, wie sie dann newlich offentlich auff sich selbs gedeutet in dem *speculo sophico*, dieses Jahr außgangen. Mögen sich derowegen hören lassen, so ihnen geliebt". Habrecht refers here to Schweighardt [= Daniel Mögling (1596–1635)] (1618)]. Schickard was on close and amicable terms with Mögling; on this, see Neumann (1995).

Michael Maestlin, Astronomischer Discurs von dem Cometen, so in Anno 1618, im Nouembri *zu erscheinen angefangen und bis im* Februario *dis 1619 Jars am Himmel noch gesehen wirt (Manuscript* WLB *Stuttgart, Cod. Math. 4° 15b, Nr. 8)*

Critical edition of the German text by Miguel Á. Granada

Michael Maestlin, Astronomical Discourse on the Comet that First Appeared in November 1618 and Can Still Be Seen in the Sky in February of this Year 1619 (Manuscript WLB *Stuttgart, Cod. Math. 4° 15b, Nr. 8)*

English translation by Patrick J. Boner
Notes by Miguel Á. Granada and Patrick J. Boner

⟨MICHAEL MAESTLIN⟩

1 [Fascicle 8a] Astronomischer Discurs von dem Cometen, so in Anno 1618, im *Nouembri* zu erscheinen angefangen und bis im *Februario* dis 1619 Jars am Himmel noch gesehen wirt.[a]

Es hatt unser getrewer Gott, zu ieder Zeit wie er gegen der welt und uns Menschen gesinnet, nicht allein durch sein h⟨eiliges⟩[b] gepredigte⟨s⟩ wort sonder auch durch mancherley Zeichen angezeigt und geoffenbaret. Also hatt Er seinem Volck, da es noch ein kleines Heufflin war, durch die der stern, in Josephs traum,[c] was Er inskünfftig vorhabens fürgehalten. Desgleichen hernach in Egypten nit nur die 7. wolfeile, sonder auch die 7. schwere und tewre Jar, mit denen auch sein Volck das haus Israel getroffen worden, durch besondere träum verkündigt. Gleichermassen hatt Er durch andere Zeichen als hie auf Erden durch Erdbiden; im waßer durch verwandlung in blut; im Lufft, mit durch seltzame *meteoris*; im Himel auch Cometen und ungewohnliche newe Stern, fürgebildet, sonderlich aber (die alten zeiten nit zu erholen) so hatt der liebe Gott nun mehr in vorigen und ietzt lauffenden *seculo* von Anno 1500. bis ietzt Anno 1619 vil und mancherlay *ostenta, portenta* und wunderzeichen an menschen und vihe, mit ongewonlichen und abschewlichen Mißgeburten, mit grausamen Waßergüssen und seltzamen Erd gewächs; mit[d] und erschrockliche fewrzeichen und *Chasmatis*: auch an himmel selbsten mit newen zuvor nie erschinen Sternen und Cometen, fürgestellt und das alles in solcher anzal, das sie nit zu erzehlen.

Bedencke man nur[e] (anders alles zu geschwaigen) wie vil Cometen in disen beeden *seculis*, von 1500. bis ietzt 1619. an den himmel sich haben sehen laßen, als in *Annis* 1500. 1506. 1510. 1516. 1522. 1523. 1527. 1529. 1531. 1532. 1533. 1538. 1539. 1541. 1542. 1545. 1549. 1556. 1558. 1560. 1569. 1572. (ein newer Stern in *Cassiopeia*) 1577. 1580. 1582. 1585. 1590. 1596. 1604. (ein newer stern in ♐ bey des *Serpentarii* füßen) 1607. und ietz 1618.

---

[a] See Figure 1 in the Introduction, page 9. Maestlin first wrote *und bis ins 1619 Jar am Himmel gesehen wurde*. Later, when still writing in February, he added *im Februario* below the line and *noch* above the line. At this point, he made the change to *dis 1619 Jars ... wirt* in accordance with the addition. See also Appendix 2 for the date 9 Febr. 1619 on p. 27, the last page in the rough copy. [b] *sein h.* added above the first writing *das gepredigte*. [c] Read *die Zeichen der stern*. Reference to Genesis 37: 5–10, more precisely the stars in the second dream. Maestlin first wrote *durch die träum Josephs*. He later added in the margin *der stern in* and changed with numbers 1 and 2 the order of *träum* and *Josephs*, without converting *träum* into the singular *traum*. [d] One illegible word follows. [e] *Added above the line*: Bedencke man nur. *Deleted*: Dan.

⟨MICHAEL MAESTLIN⟩

[Fascicle 8a] Astronomical Discourse on the Comet that First Appeared in November 1618 and Can Still Be Seen in the Sky in February of this Year 1619 1

What our true God has willed at any time for the world and us men, He has revealed and shown not only through His h⟨oly⟩ preached word, but also through various signs. Thus, He held up to His people, since they were still a small heap, what He had in store for them through the [signs] of the stars in the dream of Joseph. In the same way, through other special dreams,[1] He later heralded in Egypt not only the seven abundant [years], but also the seven difficult and disastrous years with which His people, the House of Israel, were struck as well. Similarly, He prefigured [events] by other signs, on earth through earthquakes, in the water by conversion into blood,[2] in the air through strange weather conditions [*meteoris*],[3] and in the heavens by comets and extraordinary new stars; but particularly in the previous and present centuries (not to take up old times), from the year 1500 to now 1619, God has introduced so many and various marvels [*Wunderzeichen*], portents, and prodigies among men and cattle, with extraordinary and dreadful monstrosities, with ferocious floods and foreign flora, with [] and fearsome fiery signs and chasms[4] in the sky, and with new and never before seen stars and comets in the heavens themselves, all in such quantity that they cannot be recounted here.

Just consider (not to mention any other example) how many comets have appeared in the sky in the past two centuries, from 1500 until now (1619), such as those in the years 1500, 1506, 1510, 1516, 1522, 1523, 1527, 1529, 1531, 1532, 1533, 1538, 1539, 1541, 1542, 1545, 1549, 1556, 1558, 1560, 1569, 1572 (a new star in Cassiopeia), 1577, 1580, 1582, 1585, 1590, 1596, 1604[5] (a new star in Sagittarius in the foot of Ophiuchus), 1607, and most recently in 1618.

---

1 Cf. Genesis 41: 1–36.

2 Exodus 7: 17 ff. (the first plague).

3 Ibid., 9: 18 ff. (the seventh plague).

4 On this kind of meteor, see note 7 below.

5 For a recent survey of sixteenth-century comets, see Kronk (1999, I, 293–330). Treatises on meteorology and comets from the period usually contained a catalogue of historical comets with their dates of appearance and presumed effects. The one in Rockenbach (1602, 205–235), a rather traditionalist author in cometary theory, coincides, apart from a few exceptions, with Maestlin's list. On this tradition of *cometographia* see Granada, Mosley, and Jardine (2014, 297–311). In the second part of his treatise, Schickard also provides a cometography from 1472 to the comets of 1618; see Schickard (1619, 126–143).

Was nun unser lieber Gott ietzigsmal fernes vorhabens, das ist Ihm bewußt, uns aber ligt ob, dise Zornzeichen (anderst sind sie nit zu nennen) nit zu verachten: dann was auff die vorige erfolget seye, ist bey meniglichen noch in frischer gedächtnus. Sonderlich aber will zu mercken sein, das nach dem ietzt in 11. Jaren kein Comet erschinen (da doch zwar so vil zeit von 1500. an on eins Cometen erscheinung nit fürüber gangen) der ietzig aber desto größer und deshalben erschröcklicher sich hatt sehen laßen: so will von nöten sein, dem lieben Gott mit flehen und betten in die Rutten zufallen, wie wir dann aus Gottes wort sorglich erinnert werden etc. Gott seye uns gnedig. Wir wöllen aber disen ietzigen letzten Cometen beschauen und in der furcht des herrn bedenken.[a]

[a] *Added in the margin*: Wir wöllen aber disen ietzigen letzten Cometen beschauen und in der furcht des herrn bedenken.

What our dear God now has in further store for us is known only to Him, yet we must not scoff at those signs of wrath (they should not be known by any other name); for what followed the foregoing still remains fresh in the memory of many. It is especially worth noting, however, that since no comet appeared over the past eleven years[6] (for so much time has not succeeded the appearance of a comet since 1500), the present one appeared all the greater and that much more frightful. Thus, it shall be necessary for us to surrender to God through sincere prayer and supplication, as we are duly reminded by His word etc. May God have mercy on us. We now wish to contemplate this most recent comet and consider it in the fear of the Lord.

6 The comet of 1607 marked the transit of Halley's comet.

2 **Caput I**

In verschinem 1618 seind mehr, als nur ain Comet erschinen

Es haben sich in verschinem 1618. Jar mancherlay *meteora*, oder ongewohnliche erscheinungen am Himmel erzaigt und sehen laßen, dann das Ich alhie geschweige die *Chasmata*, die hellinen, deren nit wenig gegen Mitternacht seind gemerckt[a] worden, wie auch der fewrflammen, welche meniglich, besonders die Nachtwächtern, oder feldschützen alhie zu Tübingen und anderstwo[b] vil mals, nicht nur schlechtlich am himel fladern,[c] sonder[d] bis hernider auf den Boden sehen fallen also[e] das wan sie hetten ein haus in Stätten oder flecken angetroffen, were es gleich so wol, als von einem brennenden Stral angezündet worden. Welches doch der gnädige Gott gnädig verhüttet hatt, dann es an den überschädlichen Brünsten, so durch unversehenem unfall aufgangen, und ettlich heuser, ja auch schier gantze Stät in brand gesteckt haben, laider mehr als zu vil ist. Sonder Ich will alhie ein astronomischen Discurs fürbringen[f] von Cometen dises Jars, als welche *inter meteora rariora* seind, und nicht so offt erscheinen, als die schießende Stern, oder die am himmel schwebende fewr flammen, welche in gar kurzer eil,[g] ja in eim Augenblicke verschwinden. Die Cometen aber bleiben ein Zeitlang; namlich (*si Plinio credendum*)[h] aufs wenigst 7. tag, vil derselben aber werden ettlich monat am himmel, wie albereit[i] auch diser ietzige 1618 geschinen.[j]

3 Und Erstlich in verschinem *Augusto* des 1618 Jars, solle, wie Ich berichtet werde, ein Comet sich am himel *sub Ursa maiore*, bey oder under dem gestirn so der große Beer genennet wirt, habe sehen laßen. Von disem Cometen hab Ich nichts fürzubringen,[k] dann er ist mir nicht ins gesicht kommen, ob zwar Ich mehrmal zu selben Zeit wegen eins scheinenden Cometen bin besprochen[l] worden, Ich auch mit allem fleis selbigem nachgesehen, hab Ich doch weder bey oder under dem großen Beeren, oder großen wagen,[m] noch irgendt am gantzen himmel einigen Cometen nicht sehen, oder mercken könden. Und

[a] *Added above the line*: gemerckt. *Deleted*: gesehen. [b] *Added in the margin*: alhie … anderstwo. [c] *Added*: fladern. *Deleted:* schwebend. [d] *Deleted*: vom himmel. [e] A definitive wording is lacking. Maestlin first wrote: sonder ~~auch vom himmel~~ hernider auf der ~~Erd gefall~~ … ~~seind gesehen worden also seind gesehen worden.~~ [f] *Added in the margin*: astronomischen Discurs fürbringen. [g] This word corrects an earlier *Zeit*. [h] *Margin*: Plinius lib. 2, cap. 25. [i] *Added below the line*: albereit. [j] *Deleted*: dieses verschine Jar geschehen. [k] *Deleted*: waiß … zu schreiben. [l] *Deleted*: angesprochen. [m] *Added in the margin*: oder großen wagen.

Chapter 1 2
More Than One Comet Appeared in the Previous Year 1618

Last year (1618), many *meteora* or strange appearances in the sky arose and became visible, so I shall not discuss here the *chasmata* [chasms],[7] whose blazing lights were observed not least around midnight, or their fiery flames, which were seen by many, especially by night watchmen or field guards here in Tübingen and often in other places, not only flickering fiercely in the sky, but also extending to the surface of the earth in such a way that, should they have encountered a house in the city or country, it would be just as if it had been inflamed by a burning ray of light. Our gracious God has graciously forbidden this, since from very harmful fires, rising by unforeseen accident, with several homes and even an entire city falling into flames, is unfortunately more than for me to say. Rather, I wish to deliver here an astronomical discourse on the comets of this year, as they are among the rarer *meteora* and appear not as often as shooting stars or the flames of fire rising in the sky that disappear in a very short time, or even in the blink of an eye. By contrast, comets remain [visible] for a while; namely (if we are to believe Pliny),[8] at least seven days, although many of them stay in the sky for several months,[9] like the one that just appeared this past year (1618).

First, in August 1618, as I have been informed, a comet should have appeared 3
in the sky under Ursa Major, by or below the constellation known as the Great Bear.[10] I have nothing to present concerning this comet, since I did not catch sight of it, even though I was told several times about a shining comet at that very moment and looked for it with all my industry; but I could not see or spy a comet, neither by or below the Great Bear or Big Dipper nor anywhere in the entire heaven. And it comes as no little surprise that this comet was not

7 Pliny (1938–1952), II, xxvi–xxvii, 97: "caeli ipsius hiatus, quod vocant chasma, fit et sanguinea species et, quo nihil terribilius mortalium timori est, incendium ad terras cadens inde" [there also occurs a yawning of the actual sky, called chasma, and also something that looks like blood, and a fire that falls from it to the earth—the most alarming possible cause of terror to mankind]. See also Aristotle (1984), *Meteorology*, I, 5, 342a 35ff. and the more extended description in Garcaeus (1568, 67^{r}–70^{r}), who classifies it under the meteors ignited in the inferior region of the air.

8 *Margin*: Plinius lib. 2, cap. 25. The distribution of the text into chapters does not coincide with the current one. Maestlin follows contemporary editions; for example, Plinius (1601–1616).

9 Pliny (1938–1952), II, xxii, 90: "brevissimum quo cernerentur spatium VII dierum adnotatum est, longissimum [C]LXXX". We correct the number following Plinius (1601–1616).

10 On this first comet (C/1618 Q1) see Kronk (1999, I, 333f.) and Kepler (1619, 177f.). Cf. also Introduction, Chapter 3.

nimpt mich nit wenig wunder, das bey uns alhie diser Comet nicht auch ist von gemeinen leutten, sonderlich von den[a] feldschützen animadvertirt worden, da doch dergleichen ongewohnliche *meteora*, frembde stern und Cometen gemeinlich zu aller erst von den gemeinen leutten gesehen, angezaigt und ausgebraitet werden. Nun hab Ich von den bauern alhie irgendt[b] weder vor noch nach erkundiget, sonder es ist allein von andern und frembden orten erschollen.

Zum Andern, Im monat *Nouembri*, hatt sich alle morgen früe umb 4. oder 3. uhr ein glantzende hellin von Mittag her zimlich hoch[c] über sich gestrecket, doch ein wenig, wie ein horn gebogen, sich erzaigt.[d] Sein farb war mehr weis als gelblecht, doch zu röttlecht ein wenig genaigt. Ob diser glantz, auch ein Comet, wie etlich wöllen, zu nennen seye, setze Ich iedem, der so ine gesehen hatt, frey, sintenmal er nicht gantz und volkommen in unsern landen erschinen
4 ist, sonder sein anfang, *caput* oder Comet stern[e] ist über unsern Horizontem |
nit herauf kommen. Möchten deswegen die völker, so sich ferner gegen Mittag, als in Africa, America oder Peru wohnen, zu finden sein, dann bey selbigen hatt diser glantz sich genug erzaigt. Meins erachtens solte dis *Meteor* vil mehr *Cornu*, κερατίας, oder[f] κερατοειδής, wie die *Meteorologi* solche *phaenomena* underschaiden, genennet werden. Dann über das, das er einem horn nit ungleich, als gegen seinen anfang zimlich schmal, gegen dem End gebogen, und bereit gen westen so hatts sich gegen die Sonnen, welche dazumal war im ♏[g] schier recht über zwerh gestellt, da hergegen alle Cometen ire *caudas* haar oder barth[h] von der Sonnen *in oppositum* werffen, denn die[i] Cometen so früe vor der ☉ hergehen wenden ire *caudas* gegen nidergang und die nach der ☉ nidergang gegen Abendt stehen, wenden sie gegen Aufgang etc. Also das *Gemma Phrisius*, und vor Ihme *Petrus Apianus* diser mainung waren, gleich wie, so die Sonn durch ein Christall scheinet, gerad auf der andern seitten gegen über im lufft durchscheinende stralen gesehen werden, also falle oder erstrecke sich[j] von des Cometen *corpore* oder stern, stracks und *secundum lineam rectam* gleichformige stralen oder schein. Und das dise mainung nit gar zu verwerffen seye, ist aus angeregten und hernach folgenden weit abzunemmen, seittenmal ja alle Cometen ire stralen, haaren oder Bart abwerts und von der Sonnen, keiner

[a] *Deleted:* Nachtwächtern oder. [b] *Deleted*: nichtzit. [c] *Added in the margin*: zimlich hoch. [d] *New addition in the margin*: doch … erzaigt. [e] *Deleted*: von welchen dise hellin als sein cauda oder Schwaiff sich erstrecket. [f] *Margin*: κερατίας, oder. [g] *Margin*: welche dazumal war im ♏. [h] *Added above the line*: haar oder bart. [i] *Added above the line*: die. *Deleted*: alle. [j] *Margin*: oder erstrecke sich.

observed around here even by commoners, particularly by the field guards, for such extraordinary *meteora*, strange stars, and comets are generally first seen, recounted, and widely reported by the common people. Now, I have not got information at all among farmers around here either before or after [the comet], but it is heard about only from other foreign lands.[11]

About the second comet, in the month of November there appeared early every morning at 3:00 or 4:00 a brightly shining light that by midday extended high above the horizon but became a bit curved like a horn.[12] Its color was more white than bleached, though it tended to be tinted a bit red. Whether this bright light was also a comet, as many wish to call it, I readily leave to those who saw it, since it did not entirely appear altogether in our land, for its beginning, head,
or comet star never emerged | above our horizon.[13] People living further toward 4
the South, such as in Africa, America, or Peru, may thus find the answer, for the luminary appeared sufficiently complete to them. In my opinion, this meteor should rather be called a horn, a horned star,[14] or κερατοειδής, as the meteorologists distinguish such phenomena. For in addition to the fact that it resembled a horn, extending utterly straight in the beginning and bending near the end, it was already situated toward the West, so toward the Sun, which at that time was in Scorpio almost directly in diagonal; for all comets cast their tails, hair, or beard in opposition to the Sun, since comets that arise so early before sunrise turn their tails toward sunset, while those that arise in the evening after sunset turn their tails toward sunrise, etc. Thus, Gemma Frisius and Petrus Apianus before him were of this opinion, namely that just as rays of light are seen shining through the air directly on the other side of a glass [lens] when the Sun shines through it, so uniform rays of light or illumination fall or extend evenly in a straight line from the body or star of a comet.[15] That this opinion is scarcely to be rejected is, according to the already suggested and the following text, to be widely accepted, for while all comets project their tails, hair, or beard away

---

11 See the contrasting report by Schickard (1619, 143–146, with a figure of its path), who was informed of it by peasants in Nürtingen and was finally able to see it twice there, not very far from Tübingen.

12 On this second comet (C/1618 V1), see Kronk (1999, I, 335–337), Kepler (1619, 178–185), Schickard (1619, 146–148, with a figure of its path). Cf. also here, Introduction, Chapter 3.

13 As clearly shown in the figure by Schickard (1619, 147).

14 κερατίας. According to Pliny a class of comets. Cf. Pliny (1938–1952), II, xxii, 90: "ceratias cornus speciem habet" [the 'horned star' has the shape of a horn].

15 This is the so-called "optical theory" of comets, first proposed by Petrus Apianus from observations of the comets of the 1530s. See Barker (1993), Kokott (1994), and Heidarzadeh (2008).

aber gegen der ⊙[a] wendet. Doch solches ist *secundum quid* zu verstehn, dann mehrerthail Cometen, die Ich observiert, haben ire *Barbas* nicht gar schier schlechtin von der Sonnen *in oppositum* gestreket, sonder umb ein zimlichs darvon abgewendet.[b] Dises *Cornu* aber, oder so es ja ein Comet soll genennet
5 werden, hatt sein haarlock[c] vil zu | fern hierüber gespreitet dan da es sonsten hette gar under Horizont mussen verborgen bleiben (seitenmal sein haupt drunder war), so hatt es[d] der 9. und. 10. *Nouembris* und ettlich tage hernach durch die *stellas Corui*, von Horizonte, gar nahendt hinauff bis zur *Ecliptica* in ♍ und nach dem 16. *Nouembris* durch die *stellas Crateris* schier bis zu des Lewens hertz gelanget. Ist[e] also *recte* binnen von der ⊙, welche zu end des ♏ oder Anfang ♐ auf dis *meteorum* gleichsam winckelrecht gefallen. Ferner, es hatt sich auch in der mitte dises horns[f] bisweilen ein zimlich große helle sehen laßen, welche vil leut für ein erschreckliche brunst hielten. Dergleiche underscheid in den Cometen nit gefunden wirt.

Dises *Cornu* neben dem, das es zwar[g] innerhalb 24. stunden, wie die andere stern, von Aufgang gegen Niedergang, seinen lauf volbracht, und wie gesagt, früe, umb 4 oder 3 uhr widerkommt, hatt es sich *motu proprio* allgemach weitter gegen nidergang (h. e. *contra signorum ordinem*) geländet.[h] Dann als Ich es erstmals gesehen, ists vom Horizonte durch die *stellas Corui* gezogen, da es on zweifel zu vor noch neher gegen Auffgang ist gestanden: In folgenden tagen ist es über *stellas Crateris*, welche *occidentaliores* seind, kommen.

6 Wanns zu leuchten hab angefangen, kann Ich nit erfahren. | Von glaubwürdigen leutten aber vernimme Ich, das sie habens den 7 *Nouembris*, schon am Himmel gesehen. Gleicher gestalt wann es verloschen sey, ist auch ongewiß. Ich zwar[i] als Ich deßen in erfahrung komen,[j] hab es vom 10. *Nouembris* alle morgen bis den 25 *Nouembris*,[k] so offt der Himmel hell war,[l] observiert, und das es täglich abnemme,[m] gemerckt. Als aber des Mons vollschein sich herzu nähert, und deswegen der Mon nit ehe, als bis es heller tag ward, undergieng, hatt er mit seinem hellern liecht dises *Cornu* gantz verdunckelt, welches auch hiezwischen verloschen, wie dann nach dem Abnemmen des Mons gantz nichts mehr davon ist gesehen worden.[n]

---

[a] *Added in the margin*: keiner aber gegen der ⊙. [b] *Deleted*: wie hirvon hernach folgen wirt. [c] *Added*: haarlock. *Deleted*: seinen Schwaiff. [d] *Added in the margin*: dan da … so hatt es. The ten lines after *geschlagen* are deleted. [e] *Deleted*: Hatt. [f] *Added*: horns. *Deleted*: hellin meteori. [g] *Deleted*: sampt andern sternen und mit dem gantzen Himmel. [h] *Deleted*: gezogen. [i] *Deleted*: habs alle morgen so. [j] *Deleted*: habe alle morgen observiert, da und das es taglich abgenommen wol gemerkt wan der himmel. [k] *Margin*: bis den 25. *Nouembris*. [l] *Deleted*: hell gwest, dann er vil mal mit wolcken trüb war. [m] *Deleted*: abgenommen und dunckeler worden. [n] The bottom half of the page is blank.

from the Sun, none of them turn toward it. Yet such is to be understood *secundum quid*, for most comets that I have observed have not extended their beards entirely in opposition to the Sun but turned quite a bit away from it. However, this horned star, or what should now be called a comet, spread its locks of hair | 5
far too broadly, for otherwise it would have remained hidden entirely beneath the horizon (since its head was below it); thus, on 9 and 10 November and several days afterward it extended from the horizon through the stars of Corvus all the way up to the ecliptic in Virgo and, after 16 November,[16] through the stars of Crater nearly to the heart of Leo. And so it stretched almost straight away from the Sun, which at the end of Scorpio or the beginning of Sagittarius fell similarly in a straight line on this meteor. Further, there also appeared occasionally in the middle of this horn a great bright light, which many people held as a terrible heat. Such a distinction is not found in comets.

This horn, along with completing its course from sunrise to sunset within twenty-four hours like the other stars and, as mentioned above, returned early, at 3:00 or 4:00 in the morning, gradually shifted by its own motion further toward sunset (i.e., against the order of the signs). For since I first saw it, it moved from the horizon through the stars of Corvus, as it was once undoubtedly still closer to sunrise. In the following days, it passed by the stars of Crater, which are further West.

I cannot tell when it first began to shine. | However, I hear from reliable 6
sources that they saw it in the sky already on 7 November.[17] In the same way, it is also uncertain when it was extinguished. As soon as I found out about it, I observed it every morning from 10 November to 25 November, as long as the sky was clear, and noted that it diminished daily. Yet as the light of the full moon approached, and the moon did not disappear until the day grew brighter, it completely obscured with its bright light the horned star, which in the meantime had also extinguished, as not a single trace of it was seen after the waning of the moon.

---

16 Recall that all dates by Maestlin refer to the old Julian calendar.

17 Kronk (1999, I, 335) and Kepler (1619, 180) give 11 November (new style; 1 November old style) as the date of the first observation. Schickard (1619, 138) first observed it on 9 November (old style).

7 Zum Dritten. Uber dis hatt der Allmächtige Gott noch ein anders scheinbares liecht, doch würcklich ein erschrocklich Zorn Zaichen an den Himmel gestellt. Nämlich einen Cometen, *Pogoniam*, welcher an lenge und größe vil andere Cometen weit öbertroffen. Des Cometen farb war zwar nit rot oder fewrig, sonder mehr blaich als weiß. Sein *haupt* oder der Cometstern war wie in mehrertheil anderer Cometen observiert worden. In der mitte ein dunckel glantzend Sternlein rings weis mit haaren umbgeben. Vom selbigen erstralet sich der bart oder *cauda*, zu anfang nit sonders brait, aber hinaus ie lenger ie braiter, zu end waren selbige Barts ie dinner und dunkeler, also das von sein nechsten End nit underscheiden künde.[a]

Wie es weiter mit disem Cometen *Astronomicè* beschaffen, soll in folgenden Capiteln gemeldet werden.

[a] From *Des Cometen farb*, an almost illegible addition in the right margin.

On the third [comet].[18] In addition, Almighty God situated yet another apparent light, but truly a terrible sign of wrath in the sky. This was a comet, or bearded star,[19] which far surpassed many other comets in length and size. The color of the comet was, in fact, not red or fiery, but more pale than white. Its head or comet star was observed like those in most other comets: a dark shiny star in the middle encircled by white hair. From here, the beard or tail blazed, not especially bright in the beginning, but longer and wider thereafter. At the end, the beard became ever darker and thinner, such that it could not be distinguished from its eventual end. 7

What can be further gathered from this comet astronomically shall be reported in the following chapters.

18 On the third comet (C/1618 W1), which was brighter and longer-lasting, see Kronk (1999, I, 338–341) and Kepler (1619, 185–196).

19 *Pogonia* denotes a comet bestowed with a tail or long beard. Cf. Aristotle (1984), *Meteorology*, I, 7, 344a 22–23; Pliny (1938–1952), II, xxii, 89; Seneca (1971–1972), I, xv, 4; Garcaeus (1568, 22^{r}). Schickard devotes a long description to this comet, covering its full course and accompanied by wonderful images; see Schickard (1619, 148–162).

II. Cap.
Von dises Cometen ersten Aufgang und Erscheinung[a]

Wan[b] diser Comet erstmals zuleuchten angefangen und erschinen seye, kan Ich geradlich[c] nit wissen. Ich hab Ine aller erst den 17. *Nouembris* gemerkt, da in ettlichen tagen zuvor und hienach der Himmel früe am morgen mit wolcken zimlich überzogen.[d] Den gemelten 17. *Nouembris* der himmel sich zu seubern anfieng und Ich nach hioben angezaigten *Cornu* schawete, hatt sich dises Cometen oblange *Cauda* herfürgethon und sehen lassen. Selbigen aber hab Ich dazumal für ein Cometen nicht erkennet, sonder vermeint, es were ein *radius diluculi* und ein schein oder vortrab des schier anbrechenden tags, bevorab dieweil selbiges mal sein *caput*, oder der Cometstern hinder den noch am himmel überigen wolcken verborgen war. Und in dem Ich meine gedanken das *Cornu* zu besehen gerichtet, ist mir in den Sinn nicht kommen, das zumal zu ainer zeit so zway starke und ungewohnliche *meteora* **sich solten sehen lassen** →***pag. sequenti.***

[Es[e] wirt zwar bey den *Historicis* gelesen, das bisweilen zu ainer zeit weren zween Cometen gestanden,[f] der aine spat, der ander früe, oder seye gleich einer auf erleschung des andern gefolgt. Aber so man die umbständ recht erweget,[g] befindet es sich klärlich, das es nit zween, sonder auf ainer zeit nur ain Comet gweßt seye, welcher dieweil er ettwas weitters von der ⊙ gegen Mitternacht sich erhebt, zu abend nach der ⊙ ist undergangen, und folgenden tagen vor der ⊙ wider aufgangen oder hatt sich zur ⊙ genähert under Ihren schein ettlich tag sich verborgen,[h] und dann auf der andern saitten seider herfürkommen. Aber

---

[a] Added in the right margin, some lines below the first lines of this second chapter, after the preceding marginal additions. See Figure 2 in the Introduction. [b] *Added in the left margin*: II Cap.; doubtless an indication that the second chapter begins here. [c] *Added above the line*: geradlich (uncertain reading). *Deleted*: aigentlich. [d] *Deleted*: erstmals den 17. *Nouembris* gemerkt, als Ich nach dem hie oben angezaigenen *Cornu* geschawet. [e] We use square brackets to distinguish the text Maestlin intended to leave out in order to continue the discussion on the following page. [f] *Deleted*: gesehen worden. [g] *Deleted*: betrachtet. [h] *Deleted*: gebliben.

---

20 Kronk (1999, I, 338) gives 25 November (UT) as the date of the first appearance, that is, two days before the observation by Maestlin (always in old style). Kepler (1619, 185 f.) observed it on 29 November; Schickard (1619, 148 f.) on 25 November (old style).

21 As the German translation says immediately afterwards, it is a ray of light at daybreak. See the Introduction, p. 5 and note 24 on Maestlin's response to Heinrich Welling's query, preserved in the manuscript codex of Johann Unfried.

22 See the indication of this on p. 7 in the rough copy (cf. Figure 2 in the Introduction), just below the addition in the right margin indicating the beginning of Chapter 2.

Chapter 2
On the First Emergence and Appearance of This Comet

When this comet first appeared and began to shine, I cannot say precisely. I only noticed it for the first time on 17 November, for the sky was quite covered with clouds in the early morning several days before and afterward. On the said day (17 November), the sky began to clear and I saw on high the horned star as indicated, the oblong tail of this comet having grown and become visible.[20] At the time, however, I did not recognize it as a comet, but supposed it was a *radius diluculi*,[21] a glint or forerunner of the freshly breaking day, since the head or star of the comet had remained hidden before that time behind the clouds still in the sky. For as I directed my attention to observing the horned star, it never occurred to me that two such great and extraordinary *meteora* should appear at the same time (proceed to the following page).[22]

[It[23] is read among the historians, in fact, that there were at times two comets that arose at the same time, one later, the other earlier, or the one immediately following the extinction of the other. Yet so long as one considers the circumstances correctly, it is clearly shown that there were not two but only one comet at a time that, since it rises slightly further in front of the Sun around midnight, has set in the evening after the Sun and in the following days risen again before the Sun or come closer to it and hidden beneath its brightness for several days, only to appear again on the other side of it. Yet here the circumstances were

23 We use square brackets to distinguish the text Maestlin intended to leave out in order to continue the discussion on the following page. Schickard (1619, 81f.; § 28: "Es seind noch nie zwen Cometen zu mahl am Himmel gestanden") is more explicit, with reference both to historians (Matteo Palmieri and Julius Cesar Scaliger) and philosophers (Aristotle, Pliny): "Hie muoß ich mich abermahl in Kampff einlassen, sonderlich mit ettlichen *Historicis*, die in ihren Cronikhen auffgezaichnet haben, daß ettwan auff eine Zeitt, zwen Cometen zu mahl erschienen seyen, deren der ein morgens vor der Sonnen Auffgang gegen Orient ubersich, der ander abends nach der Sonnen Nidergang, gegen Occident undersich oder auff die Seitten hinauß gestanden sey. *Matthaeus Palmerius* ein Florentiner sezt *Anno Christi* 729. seyen zwen solcher gwest; deren einer vor der Sonnen, der ander hinder ihr gestanden sey. Ja der hochgelehrte *Julius Caesar Scaliger* ist selbsten hierinn betrogen worden, der *Exerc.* 70. schreibt, er hab mit seinen eignen Augen, vil Tag lang nacheinander, zwen solcher Cometen zu mahl gesehen, einen *matutinum*, den andern *vespertinum*. Welches under den alten *authoribus* der *Aristoteles* auch für müglich helt, 1. *Meteor*. 6. Aber der *Plinius lib.* 2. *cap.* 25 nit unbillig hieran zwaiffelt, und sagt: *nemini compertum esse alteri, quod ipse quidem sciat, praeterquam Aristoteli.* Eben diser *rumor* und falsches Geschray ist auch von dem letst erschienen Cometen Flugs under gemeinen Leutten spargirt und allenthalben außgebraittet worden".

8 alhie hatt sichs recht | anderst befunden dann zur selben Zeit ist das *Cornu*, (darvon droben)[a] so noch in mercklicher[b] größe gegen mittag werts, der neue aber gegen auffgang erschinen.[c] Da dann aines hatt für das ander nit könden angesehen[d] werden.]

Das[e] aber selbiger glantz des Cometen *cauda* geweßt seye, ist daraus offenbar, das die rechte tagshelle *aurorae principium* wider mein verhoffen, nit hatt[f] hernach folgen wöllen, sonder dieweil es noch zu früe (umb 4. uhr) vil langsamer kommen ist. Es hatte auch anders, der *proprius motus huius Cometae* in folgenden tag zu erkennen geben, das eben den 17 *Nouemb.* der Comet am selbigen ort, da der sein schein hergestigen war gestanden sei.[g]

[Fascicle 8c, p. 9; see Figure 4] Sirius genandt, wirt jährlich zu end des Aprilen, zu abendt nach der Sonnen Undergang nicht mehr gesehen, dann sie kompt im zu nahe, und verdunckelt in. Alda bleibt er verborgen bis schier mitten im Augstmonat, da die Sonn für in so weit ist geschritten, das er früe am Morgen sich vor der Sonnen Aufgang wider sehen laßt. Es bereden sich die Kinder (damit Ich solchs auch melde) und zwar ettwan auch Alte Leut, sie sehen den Hundts stern in den Hundts tagen in eim Spiegel im waßer, etc. Ist aber ein eitelige fantasey. Dann welches daselbsten gesehen wirt, ist kein Stern, sonder ein widerschein im Spiegel von der Sonnen, und wirt nicht nur in Hundts tagen, sonder durchs gantze Jar gesehen, etc.

Aber alhie[h] mit disem Horn und Cometen war es weit anderst beschaffen. Dorten kundt es geschehen, das ein Comet für zween ward gehalten, dieweil er zu underschidlichen Zeitten sich sehen ließ, hie aber als das Horn noch in mercklicher größe war, und gegen Mittag werts stund, eben zu selbigen Zeit hatt sich auch der Comet gegen Aufgang erzaigt, waren also ettlich tag[i] zumal beede mit einander zu einer Zeit und stund gesehen. Darumb kundt man nit eines für das ander halten, oder ains für zway ausruffen.

---

[a] *Deleted*: gehandlet worden) und dises welches. Maestlin refers to the second comet. [b] *Deleted*: zimlicher. [c] *Added in the margin*: gegen mittag werts, der neue aber gegen auffgang. Deleted: helle, und diser Comet an einem andern besondern ort erschinen. [d] *Deleted*: und gehalten. [e] *Marginal note*: ⊗ pag. praeced. See note 24 in the translation. [f] *Added in the margin*: aurorae principium wider mein verhoffen, nit hatt. [g] *Added in the margin*: Es hatte … gestanden sei. Our reading of this addition is uncertain. [h] Cf. the parallel text on pp. 8–9 in the rough copy. [i] *Margin*: ettlich tag.

found to be quite different |, for at the same time the horned star (mentioned 8
above) was still of a notable size near the meridian, the new [comet] appeared around sunrise. Thus, one could not be regarded as the other.]

That[24] it was the same light from the tail of the comet is evident from the fact that the beginning of dawn (against my hope) did not intend to follow afterward, but since it was still too early (around 4:00) came far more slowly. In addition, it was possible to discern from the proper motion of this comet the following day that the comet was still on 17 November in the same place, for it had begun to grow brighter.

[Fascicle 8c, p. 9; see Figure 4] The star called Sirius is no longer seen in the evening after sunset near the end of April each year, for the Sun comes too close to it and obscures it. Then, it remains hidden [by the Sun] until almost the middle of August, when the Sun has proceeded forward so far that it can be seen again early in the morning before sunrise. As children say (so that I may also report it) and even the elderly at times, they see the Dog Star in the dog days[25] of summer in the water with a mirror. But it is an empty fantasy. For what is actually seen is not a star, but a reflection in the mirror of the Sun, and it is seen not only in the dog days [of summer], but throughout the entire year.

Yet things turned out far differently in the case of the horned star and the comet.[26] For in other cases, it could happen that one comet was reckoned as two since it was seen at different times, but here the horned star was still of a notable size and situated toward the South when, at the very same time, the comet also appeared toward sunrise, and both were seen together at the same moment in time for several days. For this reason, no one could tell one apart from the other or proclaim one as two.

---

24 *Marginal note*: ⊗ pag. praeced. Maestlin indicates that the text connects with the preceding page, skipping the bracketed passage. It is most probably that the transition in the clean copy (fascicle 8c) from the first sheet to the second was made here, that is, from the lost page 8 to page 9, the first one preserved in the clean copy. Accordingly, we continue (after this paragraph from p. 8 in the rough copy or fascicle 8a) the edition with the transcription of the clean copy. The reader should be aware that a strict continuity is absent; the first paragraph in the clean copy (p. 9: "Sirius genandt ...") reveals a new case of misinterpretation of celestial appearances, whereas the second paragraph ("Aber alhie mit disem Horn und Cometen ...") connects directly with the text on pp. 7–8 in the rough copy.

25 The dog days of summer, following the heliacal rising of Sirius after its cosmical setting.

26 Maestlin means the second (horn) and third comets. He accepted that both coincided in the sky for some days, but doubted whether the horn was a comet in the heavens and preferred to call it a *meteor*, the traditional term for a sublunary manifestation. Cf. below, p. 11, the conclusion of this second chapter.

Es laßen sich ettliche vernemmen, sie haben disen Cometen bey 8. oder 10. tag zuvor (vor dem 17. *Nou.*) gesehen. Nun haltens die *Astrologi* darfür, das die Cometen gemeinlich iren anfang haben zur Zeit eins Newmons. Und also rechnet Haly ein Araber,[a] und zu seiner Zeit ein berhümter *Astrologus*, den Anfang der Erscheinung eins Cometen, welchen Er in seiner jugent observiert, auf den newen Monschein, welcher (wie das *thema coeli, et positus omnium Planetarum calculo Astronomico* examiniert ausweiset) war der letzte *Aprilis, Anni Christi* 1006. Nach solcher der *Astrologorum* Regel möchte vermutlich sein, es seye diser Comet mit dem newen Mon den | 7. *Nouemb.* umb 7. Uhr vor Mittag, angezündt worden. Diser mainung will Ich zwar auch nit widersprechen, dann das es also möchte geschehen sein, ist wol müglich. Aber selbigen Tag (das ist, 10. tag vor dem 17. *Nouemb.*) wie auch ettlich tag her nach, war es unmüglich, das ine am morgen jemandt hette sehen könden. Dann nach ausweisung der proportion und *analogiae* seins lauffs, ist Er, wie hernach wirt angezaigt werden, nicht vor der Sonnen hergangen. Die Sonn war im 25. grad des Scorpions, der Comet aber (welcher one Zweifel in selbigen tagen eben die proportion in seinem lauff, wie hernach, gehabt) were nit weit von den letzten graden des Scorpions, und zimlich weit in *austro*, gestanden, hette also der Sonnen nachgefolget. Were deswegen nicht in vnsern Landen, sonder von den Völckern gegen Mittag, in America, Peru, und bey den *Antipodibus*, wie auch in *exterioribus partibus Africae*, gesehen worden. Oder wann Er ie auch in unsern Landen sich erzaigt hette, so were es zu Abendt nach der Sonnen Undergang geschehen. Welches auch schier nit aller dings zu vernichten scheinet. Dann Ich werde seidther zum thail von andern orten schrifftlich berichtet, zum thail thun ettliche in iren von disem Cometen ausgegangnen Schrifften austruckliche Andeuttung, das zuvor, ehe diser Comet am morgen erschinen ist, an ettlich orten am Abendt nach der Sonnen Nidergang auch ein Comet sich erzaigt habe.

Welchen tag aber Er sich von der Sonnen früe habe herfür gethon, ist mir auch unbewußt. Doch ists *ex analogia motus* in allweg vermuttlich, sein *caput* seye vor dem 16. *Nouemb.* aus der Sonnen strahlen und schein nit kommen, doch möchte sein Bart, ainen, villeicht zween tag, aber gewißlich nit zeittlicher zuvor, gesehen worden sein, wie er auch den 17 *Nouemb.* noch dinner und dunckeler war, als er hernach worden.

---

[a] *Margin*: Haly Comment. in cap. 9. Quadrip. Ptolemaei.

---

27 The reference in the margin (see note to the German text) is to Chapter 9 of the second book of the *Quadripartitum opus cum commento Albohazen Haly*, Venice, Octavianus Scotus, 1493, 46ᵛ–47ʳ. The actual author of the commentary was the Egyptian physician and astrologer ʿAlī ibn Riḍwān (ca. 988–ca. 1061), but the printed Latin translation from 1493

There are many who say they saw this comet eight or ten days earlier (before 17 November). Now the astrologers hold that comets generally have their beginning at the time of a new moon. Thus, Haly,[27] an Arab and in his own time a famous astrologer, calculates the beginning of the appearance of a comet that he observed in his youth according to the new moon, which was the last day of April 1006, as the figure of the heavens [*thema coeli*] and the positions of all the planets according to astronomical calculation show.[28] According to such a rule of the astrologers, it may be presumed that this comet was kindled with the new moon on | 7 November at 7:00 in the morning. I do not even intend to contradict this opinion, for it is quite possible that it could have happened this way. However, on the same day (that is, ten days before 17 November), as well as several days afterward, it was impossible for anyone to have seen it [the comet] in the morning. For according to the demonstration of the proportion and rate of its course, it did not proceed before the Sun, as shall be shown below. When the Sun was in 25° Scorpio, the comet (which on those days undoubtedly maintained the same proportion in its course, as it did later) would not have stood far from the final degrees of Scorpio, rather far to the South, and thus would have followed behind the Sun. For this reason, it would not have been seen in our lands, but by the people living around the equator in America, Peru, and at the antipodes, as well as in the outer parts of Africa. Or when it would have also appeared in our lands, it should have happened in the evening after sunset. However, this does not seem to be entirely without sense. For I am told since that time, partly in letters from other places, partly from the many texts published by others about this comet that make public notice [of it], that before this comet appeared in the morning a comet also appeared in many places in the evening after sunset.

I also do not know which day it first surfaced from the Sun. Yet it is entirely possible from the rate of motion [*ex analogia motus*] that, although the head of the comet did not emerge from the bright rays of the Sun before 16 November, its beard might have been seen before by one, perhaps two days, but certainly not any time before then, as it was even darker and thinner by 17 November than it later became.

---

attributed it to the contemporary Tunisian astrologer Abû l-Hasan 'Alî ibn Abî l-Rijâl (Albohazen Haly filii Abenragel).

28 Maestlin was possibly indebted to Cardanus (1554, 156 ff.), who first called the 'spectacle' a comet and added the astrological "Thema Cometae Haly" (p. 157), with the positions of all the planets. Still described as a comet by Pingré (1783–1784, 363–365), it is now considered to be the explosion of the supernova of 1006, the most brilliant witnessed in historical times. See Goldstein (1965), Stephenson, Clark, Crawford (1977).

11 Aus erzehltem ist leuchtlich abzunemmen, das bey den jenigen, welche sich selbs bereden, sie seien diß Cometen ettlich vil tag zuvor (doch am morgen) gewahr worden, ein Ihrtumb fürlauffe. Dann was sie gesehen haben, hatt mit nichten der Comet sein könden, sonder es ist das andere dazumal erschine *Meteorum*, nämlich das *Cornu* geweßt. Als aber der Comet auch erfolget ist, und sie nicht geachtet, das auf ein mal zway *meteora* vorhanden waren, haben Sie gemeint, selbiges *Cornu* sey eben diser Comet geweßt.

From what has been said above, it is easy to see that an error occurs among 11
those who tell themselves they first became aware of this comet several days before (but in the morning). For what they saw could not have been the comet, but rather the other meteor visible at the time, namely the horned star. Yet when the comet also arose and they did not notice that there were two *meteora* present in the sky at the same time, they thought the same horned star was also the comet.[29]

29 Cf. the text, almost identical, in the rough copy, pp. 8–9: "Daraus gilt abzunemmen, das bey denjenigen, welche meinen, sie seyen dises Cometen ettlich tag zuvor [p. 9] gewar worden ein Irthumb fürlauffet, das sie nämlich zuvor das hieoben gesetzte *Cornu*, welches auch nit fern von der ⊙ gestanden, gesehen und hernach vermeint, selbiger were eben diser Comett, da doch es zway underschidliche *meteora* waren; und in folgenden tagen beede zumal auch an underschidlichen orten am himmel gestanden, gesehen werden". Next, after a horizontal line from left to right separating the preceding text from what follows, the rough copy presents a new section with the title (in the left margin) "An Cometae prognosci possent?", which continues until the end of p. 10 and whose transcription we offer in the Appendix 1. Finally, p. 11 in the rough copy begins with a new section, entitled in the left margin "Motus via", whose beginning clearly coincides with the start of the third chapter in the clean copy: "In seinem aignen lauff hatt sich diser Comet ettlich andern Cometen fast gleich gehalten, erstlich schnell, hernach langsamer, letzlich so gmach, als wol er still stehen, *stationarius* sein, oder zu ruck sich wenden, *retrogradus* wenden". Thus, from this point, the fascicle a offers a first draft of the final writing in the fascicle c, with the exception of p. 27, where we find (as indicated in our Preface) the draft of Maestlin's apology to the Duke for having not yet completed the requested report on the comet (see also Appendix 2 for the transcription of this apology).

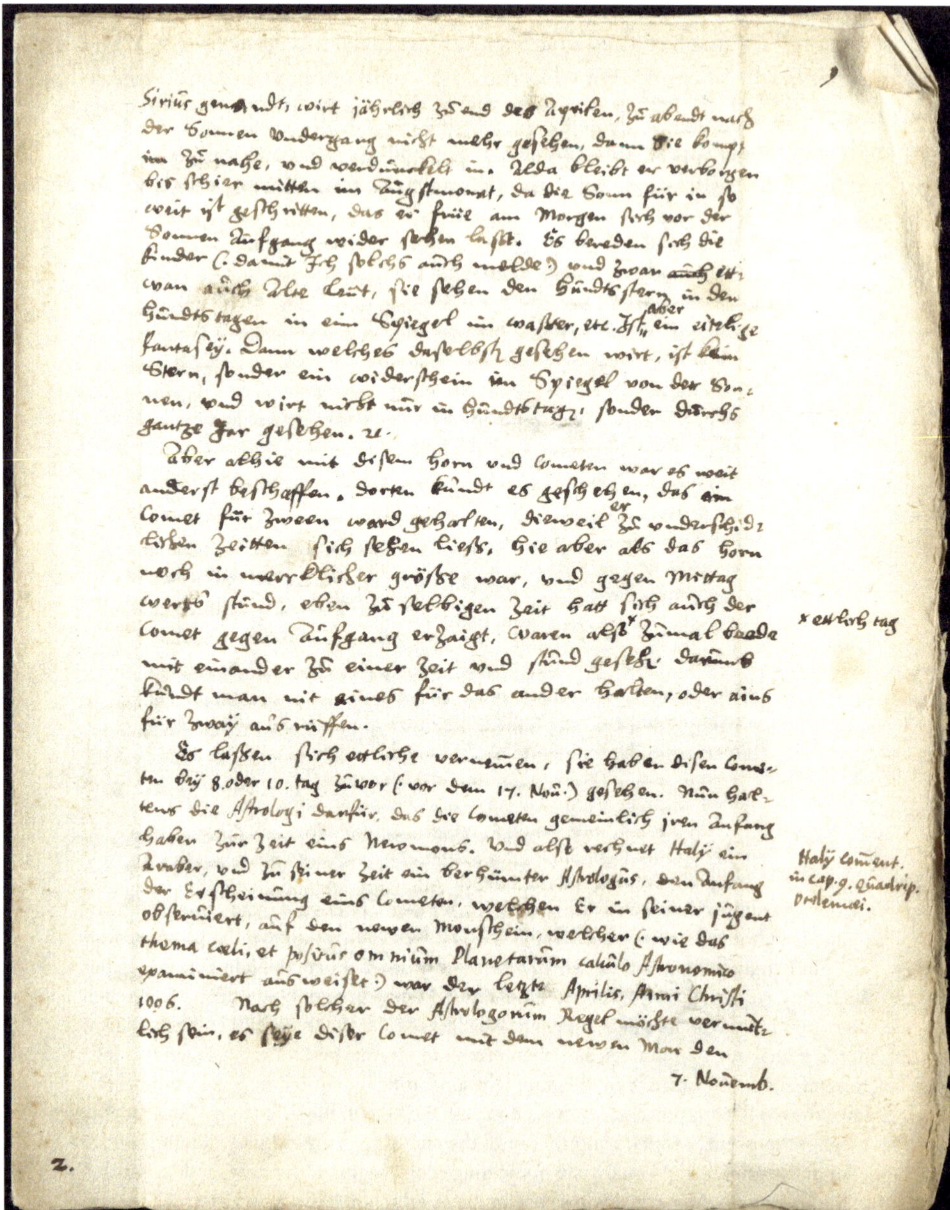

Sirius genannt, wirt jährlichs zu end des Aprillen, zu abendt nach
der Sonnen Undergang nicht mehr gesehen, denn die Sonn ist
im zu nahe, und verdunckelt in. Alda bleibt er verborgen
bis schier mitten im Augstmonat, da die Sonn für in so
weit ist geschritten, das er frue am Morgen sich vor der
Sonnen Aufgang wider sehen lasst. Es bereden sich die
Kinder (damit Ich solchs auch melde) und zwar ett-
wan auch Alte Leut, sie sehen den Hundtsstern in den
Hundtstagen in ein Spiegel im wasser, etc. Ist aber ein eitelige
fantasey. Dann welches daselbsten gesehen wirt, ist kein
Stern, sonder ein widerschein im Spiegel von der Son-
nen, und wirt nicht nur in Hundtstagen, sonder durchs
gantze Jar gesehen. etc.

Aber allhie mit disem Horn und Cometen war es weit
anderst beschaffen. Dorten kundt es geschehen, das ein
Comet für zween ward gehalten, dieweil er zu underschid-
lichen Zeitten sich sehen liess. Hie aber als das Horn
noch in merrklicher grösse war, und gegen Mittag
werts stund, eben zu selbigen Zeit hatt sich auch der
Comet gegen Aufgang erzaigt, waren also[x] zumal beede — x etlich tag
mit einander zu einer zeit und stund gesehen, darumb
kundt man nit eines für das ander halten, oder eins
für zway aussruffen.

So lassen sich etliche vernemen, sie haben disen Come-
ten bey 8. oder 10. tag zuvor (vor dem 17. Nov.) gesehen. Nun hal-
tens die Astrologi dafür, das die Cometen gemeinlich jren Anfang
haben zur Zeit eins Newmons. Und also verhaist Haly ein — Haly comment. in cap. 9. Quadrip. Ptolemaei.
Araber, und zu seiner Zeit ein berhumter Astrologus, den Anfang
der Erscheinung eins Cometen, welchen Er in seiner jugent
observiert, auf den newen Monschein, welcher (wie das
thema coeli, et positus omnium Planetarum calculo Astronomico
examiniert ausweiset) war der letzte Aprilis, Anni Christi
1006. Nach solcher der Astrologorum Regel möchte vermut-
lich sein, es seye diser Comet mit dem newen Mon den
7. Novemb.

2.

FIGURE 4 First page of the clean copy (fascicle 8c)

FIGURE 5 W. Schickard, *Cometen Beschreibung*, WLB, Cod. math. qt. 43, fol. 157, showing the path of the third comet of 1618.

CAPUT III

Von dis Cometen Lauff, welche Zaichen, und himlische Bilder Er durchwandert seye

Es hatt sich diser Comet *motu proprio* gutter maßen ettlich andern Cometen gleich gehalten, in dem er (neben *motu diurno*, da er in 24. stunden, wie anders Gestirn, umb die gantze Erden geloffen, und sich aller Welt erzaigt) in seinem aignen gang, erstlich nit so gar schnell geloffen, aber von tag zu tag schneller worden, hernach aber alle tag langsamer geweßt, bis er zuletzt schier gar still stehen wollen, und *stationarius* zu werden das Ansehen bekommen.

Sein Weg oder Straß[a] war von Mittag gegen Mittnacht, doch nit schnurschlechts, sonder beyseits gegen Nidergang (*contra signorum seriem*) sich ziehendt. Die Zaichen, die er durch gangen, sind, aus dem Scorpion, durch die Wag, die Jungfrawen und Lewen, bis in das Zaichen des Krebs. In disem seinem gang ist er under einem gewißen *Circulo magno*, bis in den Jenner dises 1619. Jars fortgezogen. Darnach aber ist er ettwas darvon abgewichen.

12 Wie sich aber diser des Cometen *circulus* gegen der *Ecliptica* halte, wirt aus den *observationibus, per calculum triangulorum* erlernet. Als, den 25. *Nouemb.* war der Comet von mir observiert, und gefunden in 3. grad. 0. scr. ♏von der *Ecliptica secundum latitudinem* 23. gr. 15. scr. Widerumb den 14 *Decemb.* war er verrückt, und kommen in den 19. gr. 50. scr. ♍, habens *latitudinem* 58. gr. 48. scr. *Calculus triangulorum* findet, das der *circulus*, oder des Cometen Straß durch die *Eclipticam* laufft im 15. gr. 23. scr. ♏, und das *angulus sectionis huius circuli et eclipticae, sive circuli cum ecliptica obliquitas* seye 63. gr. 28. scr. Under diesem *circulo* ist der Comet *praecisè* gebliben, wie Ich dann ine in allen *observationibus* durch den *Nouembrem* und *Decembrem* auf diser Straß hab gefunden. Aber im *Ianuario* dises 1619. Jars, hatt er seinen gang geändert, ist von diser Straß abgewichen, und so weit von der *Ecliptica* nit komen, als gemeldte sein Straß ausgewisen. Dann den 6. *Ianuarii* hab Ich ine gefunden im 4. gr. 36. scr. ♌ *cum latitudine* 57. gr. 10. scr. Were er aber in seiner Straß gebliben, so were sein *latitudo* gweßt 63. gr. 3. scr. Also den 13. *Ianuarii* fand ich in *per observationes* in 27. gr. 50. scr. ♋, von der *ecliptica* 57. gr. 27 scr. Da doch sein *latitudo* hatt sein sollen 62 gr. 22 scr. alles auf 5 Uhr vor mittag gerechnet.

Seines Lauffs *analogicè* nach zu rechnen (wan anderst diser Comet solte ettwas zeittlichers entzündt sein, und von selbiger Zeit sein *motum* wie hernach

---

[a] *Margin*: *Via Cometae.*

---

30 In actuality, Maestlin had been a Copernican since the 1570s. See Jarrell (1971), Westman (1972), Granada (2007, 2013).

Chapter 3
On the Course of This Comet, and the Signs and Constellations through Which it Passed

This comet maintained a proper motion in much the same way as many other comets, in which it (in addition to the diurnal motion, as it traveled around the entire earth and appeared to the world every twenty-four hours, like every other star)[30] proceeded on its own path, at first not so fast, but growing swifter every day before then becoming ever slower, until it finally appeared to stand almost entirely still and become stationary.

Its course or path[31] was from South to North, though not moving straightforward but, on the contrary, alongside sunset (against the series of signs). The signs through which it passed start with Scorpio, then Libra, Virgo and Leo, all the way to the sign of Cancer. In this way, it proceeded on a particular great circle [*circulo magno*] until January of the present year (1619). It then began to deviate slightly from this [course].

How this circle of the comet was oriented toward the ecliptic shall be found 12
from the observations by triangulation. When I observed the comet on 25 November, it was found to be at 3°0′ Scorpio, according to a [northern] latitude of 23°15′ from the ecliptic. By 14 December, it had shifted to 19°50′ Virgo, possessing a latitude of 58°48′. According to triangulation, the circle or course of the comet passes through the ecliptic at 15°23′ Scorpio,[32] and the angle of the section of this circle and the ecliptic, or the obliquity of the circle with the ecliptic, is 63°28′. The comet remained precisely on this circle, as I then found it on this same path in all [my] observations through November and December. However, it changed course in January of the present year (1619) and deviated from this path, proceeding not as far from the ecliptic as the said path had indicated. For on 6 January I found it at 4°36′ Leo with a latitude of 57°10′. Had it remained on its course, its latitude would have been 63°3′. Then, on 13 January I found it by observation at 27°50′ Cancer, 57°27′ from the ecliptic. Yet at this point its latitude should have been 62°22′, everything having been calculated for 5:00 in the morning.

If we calculate the course of the comet proportionately (that is to say, if this comet should have been kindled some time before and subsequently main-

---

31 *Margin*: *Via Cometae*. For the course of the comet, see also Kepler (1619, 196–198).

32 Cf. p. 13 below: “den 17. *Nouembr.* über die *eclipticam* im 15. grad. ♏ gestigen”. Kepler (1619, 197) gives 16° ♏.

in gleichformiger proportion ghabt haben) so ist zu vermutten, er habe sich anfänglich under des Scorpions bild, gegen Mittag, bey dem Gestirn, so von
13 den | alten *Astronomis* genennt wirt, *Bestia, Fera, Lupus*, das Gwild, oder wolf, welches der *Centaurus* stichet, erzaiget, und also in den Mittägigen Ländern, und bey den *Antipodibus* wie hieoben anregung geschehen, sich sehen laßen. Hernach were er den 17. *Nouembr.* über die *eclipticam* im 15. grad. ♏ gestigen, und zu den Wag schüßeln kommen, den 19. *Nouembr.* war er nit fern von der obern Wagschissel, *prope lancem boream* ♎, im 13. gr. ♏, *habens latitudinem boream* 5. gr. 40. scr. Den 27. *Novemb.* war er nahendt bey eim Sternlin, *tertiae magnit.* auf des *Bootis* oder Wagenmans rechtem Schinbain (*numero* 19.). Alda faßet er *in cauda* den *Arcturum*, welcher stern nach dem grossen Hundesstern, *Sirio*, under allen fixsternen der größest ist. Von dannen ist er hinauf gestigen durch den *Bootem*, raiset den 3. *Decemb.* bey den zwaien Sternlin[a] auf des *Bootis* rucken (num. 17. und 18.) durch, bis das er zu des großen Wagen vorderm Pferdt kam, da[b] stellet er sich den 12. und 13. *Decemb.* zwischen selbiges Pferdt und des *Bootis* lincke Hand. Folgendts ist er zwischen dem großen und kleinen Wagen, uber des Trachen Schwantz gestigen, und den 19. *Decemb.* bey deßelben zwaien sternlin (num. 28 und 29.) gestanden. Bisher war sein gang richtig under seinem *circulo*. Aber bald in folgenden tagen hatt er sich gewendet, und zur lincken abgetretten, sich gemählich zu des großen Beeren Haupt gelenckt, bis er letztlich nit fern von seinem lincken Ohr (stella num. 6.) vil tag im *Februario* bis in *Martium, velut stationarius*, gleichsam er still stiende, hatt zugebracht.

In seinem lauf ist er zu anfang, wie gemeldet, nicht so gar schnell geweßt, aber bis schier in die mitten des *Decembris* hatt er, wie an der größe zugenom-
14 men, also im lauff schneller fortgerücket, dorten täglich bey | zweyen graden, aber darnach über drey grad geschritten. Aber nachgehendts zumal an der größe abgenommen, und im lauf langsam worden, bis das sein *motus* anderst nit gemerckt ward, als das er nach ettlich tagen gegen den andern Sternen sich geändert hatt, ist auch ettlich wochen nur wie ein *obscura et nebulosa stella*, ein kleines dunckels Nebelin, kaum gespiret worden.

---

[a] *In the ms.*: Sterlin. [b] *Added in the margin*: kam, da.

---

33 Cf. Chapter 2, p. 10 (clean copy).

34 Ptolemy (1984, 347), Book VII, Chapter 5 (Star Catalogue): star 19 "on the right heel" of Bootes (ζ in modern designation). Cf. the path of the comet in the beautiful aquarelle in Schickard (1619, 157, Figure 5), and see Introduction, p. 5 and note 24 on Maestlin's statement of 1 December: "Cometa hic versatur circa Arcturum".

tained its motion from that moment according to the same proportion), we may suppose that it first appeared around midday under the sign of Scorpio, near the constellation | known by ancient astronomers as *Bestia, Fera, Lupus,* 13
the Wild Beast or Wolf, which Centaurus stabs; and thus it became visible to countries around the equator and at the antipodes, as suggested above.[33] Then, on 17 November it would have risen above the ecliptic at 15° Scorpio and arrived at the pans of Libra, on 19 November not far from the upper pan, near the northern pan of Libra, in 13° Scorpio, possessing a northern latitude of 5°40′. On 27 November, it was near a small star of the third magnitude in the right shin of Boötes or the Wagoner (no. 19).[34] Its tail then extended to Arcturus, which after the great Dog Star, Sirius, is the greatest of all the fixed stars. From there, it ascended through Boötes, passing by the two little stars in the back of Boötes (nos. 17 and 18)[35] on 3 December, until it arrived at the front horse of the Big Wagon,[36] where it stood on 12 and 13 December between the same horse and the left hand of Boötes.[37] Next, it climbed between the Big and Little Wagon, upon the tail of the Dragon, and on 19 December stood near the two small stars of the Dragon (nos. 28 and 29).[38] Until then, the comet had proceeded precisely on its circular path. But it soon changed course in the following days, turning to the left and advancing slowly toward the head of the Great Bear, until it finally spent several days from February until March not far from its left ear (star no. 6),[39] as if stationary or standing still.

At first, the comet did not proceed on its path so swiftly, as mentioned above, but by about the middle of December, just as it grew in size, so it began to progress more quickly, first advancing daily by | 2°, but then by more than 3°. 14
However, as it slowed it lost speed at the same pace it grew smaller, until its motion was no longer noted, except for how it had changed with respect to the other stars over several days; for several weeks, it could barely be discerned as but a dim and vaporous star, a small, dark nebula.

---

35 Ptolemy (1984, 347), σ Boo and ρ Boo in modern designation.

36 Ptolemy (1984, 343): star 27 in Ursa Major ("On the end of the tail"), η in modern designation.

37 Ptolemy (1984, 346), ι Boo in modern designation.

38 Ptolemy (1984, 345), Constellation of Draco, 10 (i) Dra and α Dra in modern designation.

39 Ptolemy (1984, 342), Constellation of Ursa Major, 24 (d) in modern designation.

Sein *caudam*[a] fand Ich nit, wie Ich an ettlich andern Cometen gesehen hab. Dann der Comet *Anni* 1577. hatt *caudam* nicht gantz *è regione* und schnurschlechts von der Sonnen gekheret, also auch der Comet *Anni* 1580. sonder sie habens vom *opposito Solis* abgewendt. Diser aber hatt *caudam praecisè in oppositum Solis* hingestrecket, on allein, das er zu end sich ein wenig zu rechten gekrümmet und gebogen hatt. Solches beweiset die *observatio* des 27. *Nouemb.* Selbigen morgen war der Comet gefunden im 29. gr. 26 scr. ♎, *habens latitudinem* 28. gr. 50. scr. Seinen *caudam* oder *barbam* erstreckt er bis ans sternlin ins großen Beeren hüfft (*numero* 22.). Dises Sternlins *locus* ist, *iuxta numeros Tychonis*, 22. gr. 48. scr. ♌, *latitudo* 35. gr. 14. scr. *Locus Solis* war 15. gr. ♐. Aus welchen *datis* findet *Doctrina Triangulorum*, das *Circulus magnus* von der Sonnen durch den Cometen, fallet in *circulum latitudinis Stellae exactè* in 35. gr. 31. scrup. eben an dis ort (ettlich *scrupula* ausgenommen, deren, weil sie so wenig, nit zu achten ist) da dis sternlin stehet. Vergleicht sich derwegen *haec caudae extensio* mit *Petri Apiani* meinung, deren droben[b] meldung geschehen. Doch hatt sich, wie erst angezaigt, *cauda* zu end ein wenig gekrümmet, aber bey weitten nit so hefftig, als wie am Cometen *Anni* 1577. gesehen worden.

Es ist auch dis an des Cometen Raiß zu mercken. In dem Er von Mittag gegen Mitternacht ist aufgestigen, ist er erstlich alle tag *supra et infra Horizontem* auf und | under gangen, doch von tag zu tag lenger über dem *Horizonte* gebliben, bis den 8. *Decemb.* da er *circulum perpetuae apparitionis* erraicht, ist er bey und alhie[c] nicht mehr undergangen, wie wol sein *cauda* ettlich tag zu vor sich der undergangs befreyet, und sich die gantze nacht hatt sehen laßen. Den 11. *Decembr.* ist er bis *in parallelum nobis verticalem* kommen, und über unser haupt alhie hergezogen. Gleicher gestalt ist er in vorgehenden tag uber denen orten, so von uns gegen Mittag, und in folgenden tagen so von uns gegen Mitternacht ligen, gezogen, so weit nämlich sein *circulus* darinn er von Anfang bis zu end seiner erscheinung seinen lauf hatt volbracht, sich erstrecket.

---

[a] *Margin*: *Caudae positus ad Solem.* [b] *Margin*: Cap. 1. [c] *Added in the margin*: bey und alhie.

---

40 *Margin*: *Caudae positus ad Solem.*

41 Maestlin (1578, 2–3): "Comam item seu caudam ab opposito Solis multum retorsit obliquo et quasi falcato tramite, quam caeteros Cometas Soli directè opposuisse refertur".

42 Maestlin (1981, xiii–xiiii): "Caudam sicut omnibus Cometis hactenus observatis comune fuit, Soli adversam, quanquam non exacte diametraliter oppositam habuit".

43 Ptolemy (1984, 342), Constellation of Ursa Major: "the star on the left knee-bend", ψ in modern designation.

I did not find its tail[40] to be the same as I have seen with so many other comets. For the comet of 1577 did not extend its tail entirely away from the Sun in a straight line,[41] nor did the comet of 1580.[42] Rather, they deviated from direct opposition to the Sun. On the contrary, the present comet stretched out its tail precisely in opposition to the Sun, except for only bending at the tip and curving slightly to the right. Such is shown by the observation from 27 November. That same morning, the comet was found at 29°26′ Libra, with a [northern] latitude of 28°50′. Its beard or tail extended all the way to the little star in the hip of the Great Bear (no. 22).[43] The location of this little star, according to the tables of Tycho, was 22°48′ Leo, with a latitude of 35°14′.[44] The location of the Sun was 15° Sagittarius. From the same data, we may find by the doctrine of triangles that a great circle from the Sun through the comet coincided with the circle of latitude of the star exactly at 35°31′, at the very same place (excluding several arc minutes, which, since they are so small, should not be considered) where this little star stood. This extension of the tail therefore agrees with the opinion of Peter Apian, which we mentioned above.[45] Yet the tail, as previously shown, turned slightly at the tip, though not nearly as sharply as was seen with the comet of 1577.

We must also remark on the journey of this comet. As the comet ascended (toward the North) from the South, it first rose above and descended below the horizon daily |, though remaining longer and longer above the horizon from one day to the next, until on 8 December, when it reached a circle of perpetual appearance, it no longer set for us, in the same way that its tail had no longer succumbed to setting several days earlier and remained visible the entire night. On 11 December, it reached a vertical parallel to our position and passed overhead. It presented the same appearance on previous days to those lands that lay to the South, and in the following days to those that lay to the North; as far, that is, as the circle therein it traversed from the beginning to the end of its appearance extended.

44 Cf. *Astronomiae instauratae progymnasmata*, TBOO, ii, p. 267: "in genu praecedentis pedum posteriorum". The longitude given by Tycho is 22° 33′.

45 *Margin*: Cap. I. Cf. the rough copy, p. 4. On Apian's observations of the comets of 1531, 1532, 1533, 1538 and 1539, see Kokott (1994, 56–72), Jerratsch (2020, 122–127) and Apianus (1540, N II^{v}–O III^{r}). For the formation of the tail by the solar rays traversing the transparent head of the comet, see Apianus (1540, N IIII^{r}): "oportet lineam a Sole per corpus Cometae euntem rectam, etiam per cometae caudam secundum longitudinem rectissimè ferri debere, adeo ex hoc omni liberatus sum dubio, imo credere coactus sum, vel invitus, caudam a Sole oriri oportere".

CAPUT IIII

Das die *Philosophi* wideriger Meinung sind, ob die Cometen *Elementares* oder *Aetherei* seien, das ist, ob sie hie unden im Lufft, oder droben im hohen himmel generiert und geschaffen werden

Ein fürneme und hauptfrag wirdt so wol bey den Gelehrten als ungelehrten fürgebracht. Bey den ungelehrten zwar mehr aus fürwütz, aber bey den Gelehrten aus rechtem philosophischem Eyfer, die *arcana Naturae*, und was in der Natur noch verborgen ist, zu erforschen, und selbige, deren noch vil seind, auch an den tag zu bringen. Namlich, wa die Cometen generiert und geschaffen werden? und wie hoch sie von der Erden erhaben seyen? Ob sie *in regione Elementari*, im Lufft, oder *sphaera ignis*, oder weitter droben im himmel ihren lauff verbringen?

Hierauff antworten der mehrerthail *Philosophi* aus dem *Aristotele*, also: „Aus der Erden und waßer, steigen täglich auf, vilerlaye gesunde und frische, vilmal aber, faule, ungesunde und gifftige Dämpf,[a] und riechen übersich, deren ettliche feucht und deswegen ettwas schweerers. Aus solchen, wan sie im Lufft resolviert werden, entstehen Regen und dergleichen etc. Welche aber subtiler und dinner, warm und trucken, die erheben sich, und fahren vil höher, wie der
16 rauch, und kommen bis in den obern Lufft, alda samlen | sie sich *in sphaera ignis*, oder nit fern darunder im Lufft, und dieweil sie ettwas faißt, kleben sie aneinander, und nachdem sie wol ausgedorret, so werden sie vom Element des fewrs, so droben über dem Lufft, doch under des Mons himmel ist, angezündet. Hieraus werden aller hand Cometen etc. Wie aber die himmlische Sphaeren täglich umb die Erden herumb lauffen, also nemmen sie, mit irem gewalt, mit sich, nit nur das Element des fewrs (als welches gar nahe ist), sonder auch den obern thail des Luffts, sampt allem was sich darinnen aufhelt. Auf dise weis werden die Cometen *motu diurno* herumb geführt. Wann sie nun angezündet seind, so wehret die brunst, bis die Materi verzehrt wirt. Letzlich was übrig, wie äschen, ist, verstiebt im Lufft, vergünfft ine, sampt wasser und Erden, und also schier die gantze Elementarische Region. Daraus dann alles Unhail entspringt, als überflüßige Hitz, oder Kälte, daher erfolgen Mißgewächs, und deswegen Thewrung; item schweere Krankheitten, Sterbendt. Also melancholisch, cholerisch Geblüt, welches verursacht widersinnisches Gemüt, und daraus nit geringe Widerwillen, Neid, Zanck, Zwittracht, Auffrhur, bis es auch zu Krieg und Bluttvergießung, verhörgung ⟨von⟩ Land und Leut kommet, etc.“.

[a] *Added in the margin*: gesunde … Dämpf.

Chapter 4
That the Philosophers Are Divided in Opinion over Whether Comets Are Elementary or Ethereal, That Is, Whether They Are Generated and Brought into Being Here Below in the Air or High Above in the Heavens

A noble and essential question is raised by the learned as well as by the unlearned. While the unlearned ask it more out of wonder, the learned are prompted by proper philosophical zeal to probe the secrets of nature [*arcana naturae*] and what still lies hidden within it, and to bring these many things to light. Namely, where are comets generated and brought into being? And how high are they raised above the Earth? Do they follow their course in the elementary region, in the air or in the sphere of fire, or higher up in the heaven?

On this, the greater part of philosophers respond according to Aristotle, saying:[46] "From the earth and water there emerge every day a wide variety of fresh and healthy exhalations, as well as filthy, foul, and poisonous ones, rising from below; many of these vapors are moist and therefore possess a heaviness. From them, when they are dispersed in the air, rain and other such things arise. On the other hand, the subtler and thinner [exhalations] that are warm and dry rise and ascend much higher, like smoke, reaching the upper air and assembling | together in the sphere of fire or not far below it in the air; and since 16
they are somewhat fatty, they adhere to one another, and after they have dried up are ignited by the element of fire, which is above the air but below the orb of the Moon. This is where all kinds of comets come from and so forth. Yet as the heavenly spheres revolve daily around the Earth, so they take with them through their violent motion not only the element of fire (as it is so close), but also the upper part of the air, including everything that lies within it. In this way, comets are led around with diurnal motion. Whenever they are kindled, the heat endures until the matter is consumed. In the end, what is left over, like ashes, spreads throughout the air, poisoning it along with the earth and water, and thus nearly the entire elementary region. Here is the source of all harm, such as superfluous heat or cold, from which a failed harvest follows and ultimately famine, as well as severe diseases and death. From it, we may trace melancholic, choleric blood, which causes a contrary mind, and with it no small measure of antipathy, envy, discord, dissent, and uproar, until it even comes to war and bloodshed, the destruction of countries and peoples, among other things."

---

46 The following long passage between quotation marks is not a direct quotation from Aris-

Wann weitter gefragt wirt, ob aber gewis seye, das die Cometen hie unden im Lufft ihre Residenz und wohnungen haben? wirt geantwort, „Dieweil die Cometen unbeständig und zergänglich seind, zu und wider abnemmen, und gantz verschwinden, so müssen sie Elementarischer Natur sein, dann solches seind aigenschafften der Elementen, und nit des Himmels, seitenmal in *aetherea regione* keine Enderungen sind, etc.“

Hergegen sehen andere vil mehr auf der Cometen Lauff, und geben auf die fürgegebne frag, dise Antwort: Es befindet sich, das die Cometen gemeinglich ein richtigen lauf, und gewissen proportionierten fortgang haben. Solche Aigenschafft aber werden den Elementen mit nichten zugemeßen. Item, wan die Cometen nur mit dem Lufft und Fewr umb die Erden herumb lieffen, so würden sie *motu proprio*, anderst nicht, als von Nidergang gegen Auffgang,
17 *secundum signorum seriem*, | gehen.[a] Dann gleich wie die planeten dem *primo Mobili*, oder öbersten Himmel (welcher sie samentlich in 24. stunden von Aufgang gegen dem Nidergang herumb führet) in Ihrem aignen lauf zu wider, von Nidergang gegen Aufgang gehen, oder wie es ettliche verstehen, ime im herumb lauffen nicht gefolgen mögen, sonder alle tag umb ettwas dahinden bleiben: Also solten auch die Cometen täglich nur von Nidergang gegen Aufgang iren gang haben, das ist, nur dahinden bleiben. Nun aber gibt die Erfahrung, das die Cometen nit alle also, sonder deren ettliche disem zu wider, *secundum signorum ordinem*, von Aufgang dem Nidergang zu ziehen, wie alberait diser unser Comet gethon hatt, und doch ein zimlich richtigen lauf behalten. Solches aber alles vergleicht sich vil mehr mit *aethereae regionis*, und des himmels aigenschafften, als mit der Elementen augenblücklichen veränderlichen unbeständigkait. Daher ettliche *Philosophi* haben die Cometen nicht nur in den himmel erhebt, sonder sie gar under die Zahl der Planeten gesetzt, und darfür gehalten, das die Cometen seien gewisse Stern, so am himmel stätigs wie andere Stern stehen, aber bleiben under der Sonnen hellem Schein verborgen, bis ir aigner *motus* sie in gewissen Zeiten laße, gleich wie zun Zeitten der *Mercurius* thut, herfür lauffen.

Under disen Zwispaltungen hatt *Aristoteles* die erste mainung, als vil scheinbarer gebillichet, dieweil sonderlich in derselbigen ettwas verständtlichers, so

---

[a] Here begins the third sheet.

---

totle, but rather a summary of contemporary Aristotelian doctrine regarding the double exhalation and the generation of comets as inflammations of the warm and dry exhalation in the superior region of the air. Cf. Martin (2011, 5–11) and for the formulation in Aristotle's *Meteorology*, see Aristotle (1984), I, 4, 341b 6–23 and 7, 344a 9–b 35.

47 Again a formulation of common Aristotelian doctrine. Cf. Aristotle (1984), *Meteorology*, I, 1–2.

Should one further ask whether we can say for certain that comets have their home and dwelling place down here in the air, they shall reply, "Since comets are inconstant and ephemeral, growing and then contracting before vanishing altogether, they must be of an elementary nature, for such are the properties of the elements and not of the heavens, as there is no change in the ethereal region and so forth."[47]

On the other hand, those who pay greater regard to the course of comets will give the following answer to the foregoing question: it is true that comets commonly possess a regular course and distinctly proportional rate of motion. However, such a property is by no means assigned to the elements. Further, if comets revolved around the Earth only with the air and fire, they would pursue their own proper motion in no other way than from West (sunset) to East
(sunrise), according to the series of the signs |. Then, just as the planets follow 17
their own course from West to East, contrary to the *primum mobile* or highest heaven (which leads them all around from East to West in twenty-four hours), or as many understand it,[48] the planets may not follow it around at the same pace, but fall behind it a bit every day, so should comets also proceed each day only from West to East, that is, only by falling behind. However, experience shows that comets, if not all, then many move the other way, according to the series of signs, from East to West, just as our present comet has now done while maintaining a highly regular course of motion.[49] Yet this all coincides far more closely with the properties of the ethereal region and the heavens than with the ephemeral and shifting inconstancy of the elements.[50] For this reason, several philosophers have not only raised comets to the heavens, but they have even located them among the number of planets, thus holding that comets may be certain stars that remain as constant in the heavens as others, though lying hidden beneath the bright light of the Sun until they proceed forward at certain moments by their own motion, just as Mercury does at times.[51]

In view of the above two alternatives, Aristotle accepted the former opinion as far more probable, for he seems to have found it particularly more reasonable

---

48 Alpetragius (al-Biṭrūjī, died in 1204) and his followers. On his reform of planetary motion, consisting mainly in an interpretation of the proper motion of the planets as an optical effect of their retarded diurnal motion from East to West, see al-Biṭrūjī (1971), Lerner (2008, I, 104–110), Sabra (1994).

49 See p. 32 below: "… diser Cometen lauff gar richtig, ordenlich und wol proportioniert gweßt sey".

50 The proportion in the motion of the 1577 comet had already been highlighted as an indication of its celestial position. See Maestlin (1578, 2, 34), Brahe (1588, 84f.), Granada (2009) for Roeslin.

51 Aristotle (1984) *Meteorology*, I, 6, 342b 29–33: "Some of the Italians called Pythagoreans

wol der Cometen Uhrsprung, herkommen, und endtlicher Undergang, vermerckt zu werden, scheinet. Und diser meinung folgen bisher der mehrer thail *philosophi*, so gar, das ettliche darfür halten, es seye diser zu widersprechen, gleichsam ein schwehre Todsünd, dann hiedurch würde nit nur die[a] *Physica Scientia*, sonder die gantze *philosophia* samentlich zu grund und boden fallen.

[a] *Deleted*: gantze.

say that the comet is one of the planets, but that it appears at great intervals of time and only rises a little above the horizon. This is the case with Mercury too". It is possible that Maestlin is indebted here to Apianus (1540, O3[r]): "Ex eo autem quod motus eius [1533 comet] fere regularis fuerit, mirari desino *quosdam* extitisse qui putaverint, Cometam esse stellam huiusmodi, quae nunc ad Solem accederet, nunc ab eodem discederet, iuxtaque Solem existens non videretur. Mercurius veluti"; our italics. Maestlin mentions Apianus on pp. 14 and 27. Cf. also Seneca (1972), 17, 1–2; 20, 4; 22 ff.

regarding the origin and provenance of comets as well as their final demise. The greater part of philosophers continue to subscribe to this opinion, so much so that many hold that opposing it would, as it were, be a grave and mortal sin, for not only would natural philosophy but the entire field of philosophy come completely crashing down from it.[52]

52 This is the opinion later maintained by Bartholomaeus Keckermann; see below, pp. 41 ff.

CAPUT V

Ob und wie in den wider wertigen Meinungen ein Außschlag zu finden seye

18 Es werden in diser disputation zu beeden Seitten, wie vermeldet worden,[a] *argumenta* auf die Baan gebracht. Aber wan man selbige alle (wafern keine bessere grund vorhanden seind)[b] erwigt, so seindts nichts, als nur *coniecturae* und muttmaßungen, welche im grund nichts gewißes[c] beweisen, sonder es sind bis daher *argumenta argumentis, ratiocinationes ratiocinationibus, coniecturae coniecturis,* eines dem andern entgegen gesetzt, welche zuletzt samentlich sovil als **nichts** gelten, dann zu beeden seitten greiffet mans nit am rechten ort, sonder hinderwerts an. Zu dem, es lauffen zu beeden theilen *praejudicia* under, da jeder thail sein *praesuppositum* für wahr helt, da doch noch keines wider das ander zu vor erwisen ist.

Wie[d] aber dise und dergleichen *Quaestiones* mögen solviert, erörtert, und der rechte Außschlag gefunden werden, lehrt Aristoteles[e] recht und wol, und zaigt an, man solle zu vörderst **das ὅτι, ob es sey**, erkundigen, darnach kan man allererst **das διότι, wie und warumb** es also oder also beschaffen seye, erforschen. Dann unmüglich ists, das einer wiße von ettwa einer sach zu reden, was und wie es darmit beschaffen seye, wann er nit zuvor wißen kan *quod sit, et an sit,* ob selbiges *in rerum natura* were oder nit? Lächerlich were es, so einer von seinem guttten freund vil erzehlen wolte, was er zu Augspurg alberait verrichte und handle, wann er noch nit wißte, ob selbiger sein freund sich daselbsten auffhielte, oder, ob er Augspurg jemals gesehen hatte, sonder müßte zwaiflen, ob er in Europa, in India oder America were. Oder, gleich wie ein Artzt zu vörderst sihet, wa der *Affectus morbi* seye, an welchem ort oder glid des leibs der patient sich übel befinde? Als dann bedencket Er, was es für ein *affectus* seye, und aus was ursachen er herkomme. Dann solte Er dem rechten Arm wöllen helfen, da doch der lincke verletzt were, so hieße es ja mehr, als den unrechten finger verbunden.[f]

Also in unser vorhabenden Materi, soll und muß man zu vorderst alle muttmaßungen hindan setzen, und aus rechten aigentlichen grund erlehrnen wa

---

[a] *Added above the line at another moment, as indicated by the different ink*: wie vermeldet worden. [b] The parenthesis was added in the margin at a later date. [c] *Added in the margin at a later date*: gewißes. [d] *Margin*: Wa man von Cometen zu reden, anfangen soll. Still a later addition. [e] *Added in the first version in the margin*: *Lib.* 2. *post. Anal. cap.* 2 et 8. [f] *Added at the bottom of the page at a later date*: Oder ... verbunden.

Chapter 5
Whether and How We May Find a Solution for the Two Opposing Opinions

There are arguments brought to bear on both sides of this disputation, as mentioned above. Yet if one considers all those things in their own right (since there are no better grounds at hand), they are nothing more than conjectures and speculations that essentially prove nothing for certain, and thus they remain arguments for other arguments, ratiocinations for other ratiocinations, and conjectures for other conjectures, each one posed against another and together amounting ultimately to **nothing**, for no one on either side grasps anything in the right place but the other way around. In addition, prejudice plagues both sides, since each one holds their own presupposition [*praesuppositum*] for certain, although neither one nor the other has ever been proven. 18

Aristotle teaches us truly how these and other similar questions might be regarded, reasoned, and resolved the right way;[53] and he shows how one should, in the first place, explore the question of ***quid*, whether it is**, before one can begin to examine the ***propter quid*, how and why** it may be so or arranged in such a way.[54] For it is impossible that someone should know how to speak about nearly anything, what it is and how it is arranged in such a way, if he cannot first know that it might be and whether it exists [*quod sit et an sit*], or whether the same thing really exists. Thus, it would be absurd if one wished to describe in detail what his good friend was now doing in Augsburg when he still did not know whether the same friend was staying there or whether he had ever seen Augsburg, but rather he would not even be sure whether his friend was in Europe, India or America. In the same way, a doctor first sees what the condition of an illness may be and in what part or place of the body the patient feels poorly before he goes on to consider what kind of condition it might be and from what causes it comes. For if he should wish to heal the right arm when the left one is actually injured, that would certainly be more than to bandage the wrong finger.[55]

Thus, in our present matter one should and must begin by setting aside all conjectures and determine from the right and proper reason where this or that

---

53 Added in the first version in the margin: Lib. 2. post. Anal. cap. 2 et 8.

54 On the distinction *οτι / διοτι* (*quia / propter quid* in the Latin translation) see also Aristotle (1984b), *Posterior Analytics*, I, 13 and Barker–Goldstein (1998, 244 ff.).

55 The additions made to this page show that the text has been carefully reworked by Maestlin, indicating its importance to his argument.

diser oder jener Comet stehe?,[a] ob auch iemals ein Comet (er habe sein uhrsprung waher es immer sey, und er gehe richtig oder nit) im himmel oder im lufft, erschinen seien? und das man solches *à priore et demonstrativè* gewiß wißen könde.

Welches ist dann der rechte und aigentliche grund, zu erkundigen wie hoch und weit ein Comet von und ob der Erden stehe? Antwort. Der rechte grund ist nit im Dampf oder rauch, so von der Erden aufriechet, noch auch fürnemlich in *regulari oder irregulari motu* zu suchen. Dann da steht es an. Es möchten die
19 Cometen iren Uhrsprung villeicht aus einer andern, uns noch unerforschten | materi[b] haben. *Item*[c] es möchten villeicht die Elementen, und was daraus herkompt, auch *regulariter* bewegt werden etc. Dann es seind in der Natur noch vil uns verborgne Heimligkaiten. Sonder sie muß zu rath gezogen werden die **Meßkunst, Geometria. Dise ist *in omni* QUANTO, *Regula veritatis*, die rechte unfählbare Richtschnur**. Dise, und ihr Schwester **Arithmetica, die Rechnenskunst**, lehren, wie die höhe, größe, weitte, ferne eins Haus, Gebäws, Thurns, Bergs, und dergleichen, ob man schon mit keiner Meßruthen, Stangen, oder Schnur kan zukomen (dann sonsten bederffte es keiner sonderen Kunst) abzumeßen seien. Und hieher gehören auch die Cometen. Hierzu braucht sie[d] keine *exhalationes*, dämpf oder rauch, sonder *instrumenta Geometrica*, und *demonstrationes geometricas*, sampt demonstrierten Rechnungen.[e] Aus welchen *tota Doctrina Triangulorum* erbawen wirt.

Ein Exempel, welches in sonderhait zu unserm Vorhaben tauget. Von einem Haus AB, so 50. schuch ist, wirt mit einem Quadranten, aus zwaien Ständen, nämlich, vom understen boden oder *basi* A, und vom Gibel B observiert die höhe eins Thurns CD. Aus dem A, zu underst am Haus, wirt die höhe des spitzes C gefunden 23. grad, aber aus dem B, zu öberst des haus, ist sie nur 20. grad.

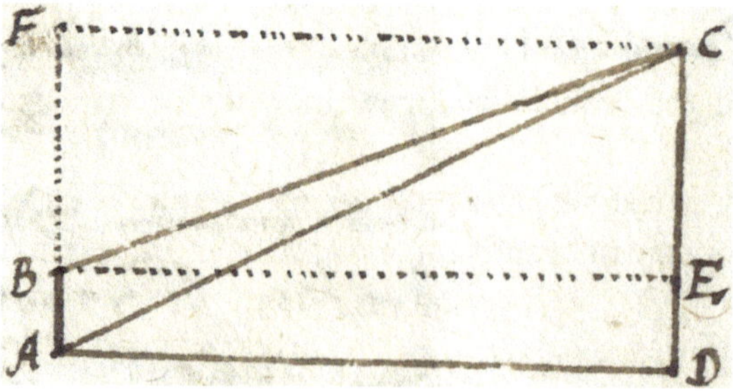

[a] *Added in the margin*: wa … stehe? [b] *Deleted*: iren Ursprung. [c] *Added above the line*: *Item*. [d] *Deleted*: *Geometria*. [e] *Added in the margin at a later date*: sampt … Rechnungen.

comet may be,[56] whether a comet has appeared in the air or in the heavens (whatever its origin may be and should it follow a regular motion or not), and that one could know for certain such things *a priori* and by demonstration [*a priori et demonstrative*].

So, what is the right and proper reason for inquiring how high and far away a comet may be removed from the Earth? Answer. The right reason is not to be found in a fume or vapor that flows from the earth, nor even principally in the regular or irregular motion [of the comet]. For that remains discussion. Perhaps comets take their origin from another source that we still have not explored |. In addition, perhaps the elements and what proceeds from them are 19
even moved in a regular motion, and so forth. For there are still many secrets in nature that remain hidden from us. However, it is **the art of measurement, geometry, which must be consulted. This [geometry] is for every quantity the standard of truth, the right and unerring rule.** Along with her sister, **arithmetic, the art of reckoning**, geometry teaches us how to measure the length, width, height, and distance of a house, building, tower, mountain, or other such thing without relying on a measuring pole, rod, or string (for otherwise there would be no need for a special art). And comets belong here as well. For them, geometry does not require exhalations, smoke, or vapor, but rather geometrical instruments and geometrical demonstrations, along with proven calculations. The entire doctrine of triangles derives from those things.

Here is an example that is particularly suitable for our present endeavor. From a house AB that is fifty feet tall, let the height of a tower CD be observed with the aid of a quadrant from two points, that is, from the base or bottom [of the house] A and from the top [of the house] B. From point A at the bottom of the house, the height of the tower may be found to be 23°, while it is only 20°

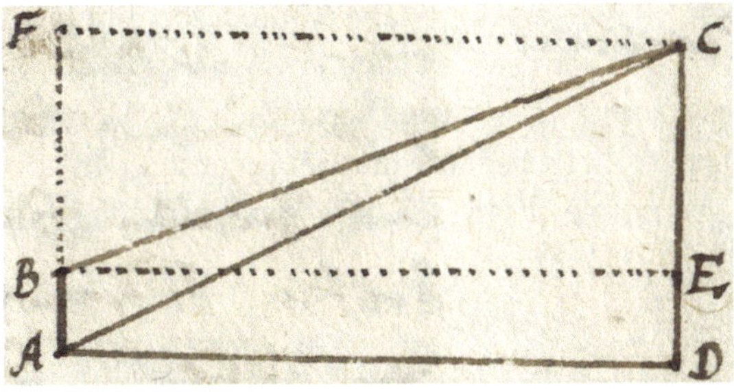

56 The same position was formulated, almost in the same terms, by Caspar Crüger in his printed treatise. Cf. Crüger (1619, 62): "Es ist wol ein wunderbar ding / das man gestritten

Wirdt gefragt, wie hoch der Thurn DC seye? Wie fern er vom haus AB stehe? *item* wie weit der Spitz C vom A, und vom B seye?

Antwort. Die *observationes* haben geben CAD *angulum* 23. grad. und CBE 20. gr. Deswegen sind ire *complementa ad angulum rectum* (nach ausweisung *doctrinae Triangulorum*) ACD, und der ime gleich ist, CAB oder CAF[a] 67. gr., also BCD, und CBF 70. gr. und auch deßen *complementum ad duos rectos* ABC 110. gr. Hiermit wirt auch bekandt *differentia angulorum* ACD und BCD, das ist, *angulus* ACB 3. gr. Jetzt lehrt *doctrina triangulorum*, wie sich helt *anguli* ACB, 3 gr. *sinus* 5233 (*ex Canone Sinuum*) gegen AB 50 schuch: also helt sich *anguli* ABC 110. gr. (oder | sein *complementum* FBC 70 gr.) *sinus* 93969 gegen AC, und *anguli* BAC 67. gr. *sinus* 92059 gegen BC. Hie bringt *Regula Detri* die Lini AC 898. schuch, und BC 879 ½ schuch. Darnach, wie sich helt *anguli* ADC *recti, sinus*, das ist, *sinus totus* 100000. gegen AC 898 schuch, also helt sich *anguli* DAC 23. gr. *sinus* 39073, gegen DC, und *anguli* ACD 67. gr. *sinus* 92050 gegen AD. *Regula Detri* bringt abermal, DC des thurns höhe 351. schuch, und die distantz des thurns vom haus, AD, 826 ½ schuch.

Es kan dise Rechnung auch also angestellt werden. Dieweil CAD, oder ACF, wie gesagt, ist 23. gr. und CBE, oder BCF ist 20. gr. so gibt *Canon Foecundus*, oder *Tangentium* die lini AF 42447 und die lini BF 36397. *Canon Hypotenusarum* aber, oder *Secantium*, gibt die lini AC 108636, und die lini BC 106418,[b] wirt also *differentia linearum* AF und BF, das ist, *linea* AB 6050 alles in der proportz darinn FC, oder AD ist *sinus totus* 100000 *partium*. So nun AB 6050 sind 50. schuch, was gibt AD? was AF oder DC? was AC? was BC? Hierauf antwortet *Regula Detri*: AD die distanz des thurns vom haus ist 826 ½ schuch. DC, der thurn ist hoch 351. schuch. AC, der spitz des thurns ist vom A, im lufft gerechnet 898. schuch. Und BC 879 ½ schuch, wie droben.

Hie ist zu mercken.[c] Diser *angulus* ACB wirt *in Astronomia* genendt PARALLAXIS. Welche nichts anders ist, als, wie erst angezaigt worden, die differentz und underschaid zwaier observationen aus zwaien Ständen oder Absehen. Hiedurch erkundigen die *Artifices Astronomi*, wie weit, sonderlich der Mon von der Erden stehe. Die Rechnung wirt aller dings, wie hieoben, angestellt. In beygesetzter figur ist A *centrum Terrae*, gleichsam der aine Stand oder das aine

---

[a] *Margin*: oder CAF. [b] *Deleted*: alles in der proportz darinn FC, oder AD ist sinus totus 100000 partium. [c] *Margin, in red*: Was Parallaxis seye.

---

und *rationes physicas* herfür gebracht / warumb die Cometen nicht können im himmel entstehen / da mans doch vor nie untersucht / Ob auch etwan einst Cometen im himmel befunden".

from point B at the top of the house. One may then ask, how tall is the tower DC? How far away is it from the house AB? Further, how far away is the top [of the tower] C from A and from B?

Answer. The observations gave angle CAD as 23° and angle CBE as 20°. Consequently, their complementary right angle (according to the demonstration of the doctrine of triangles) ACD and, what amounts to the same, CAB or CAF, is 67°; so, BCD and CBF are 70°, and their supplementary angle ABC is 110°. Hereby we may also know the difference of angles ACD and BCD, that is, angle ACB, as 3°. Now, the doctrine of triangles shows how the sine of angle ACB 3° measures 5233 (according to the law of sines) for AB 50 feet; so, the sine of angle ABC 110° (or | the supplementary angle FBC 70°) measures 93969 for AC, and the sine of angle BAC 67° measures 92059 for BC. According to the rule of three,[57] line AC is 898 feet long and line BC is 879 ½ feet long. Then, since the sine of right angle ADC, that is, the total sine, measures 100000 for AC 898 feet, the sine of angle DAC 23° will measure 39073 for line DC and the sine of angle ACD 67° will measure 92050 for line AD. Relying once again on the rule of three [*regula detri*], the height of the tower DC is 351 feet and the distance of the tower from the house AD is 826 ½ feet.

This calculation can be made in the following way as well. Since CAD or ACF, as said above, is 23° and CBE or BCF is 20°, the *canon foecundus* or table of tangents gives line AF as 42447 and line BF as 36397. Yet the *canon hypotenusarum* or canon of secants gives line AC as 108636 and line BC as 106418, so the difference between lines AF and BF, that is, line AB, is 6050, all in proportion to FC or AD, that is, the total sine of 100000. Thus, if AB 6050 is now 50 feet, what is AD? What is AF or DC? What is AC? What is BC? We may find our answers in the rule of three [*regular detri*]: AD, the distance of the tower from the house, is 826 ½ feet. DC, the tower, is 351 feet high. AC, the top of the tower from point A, measures 898 feet through the air. Finally, BC is 879 ½ feet, as mentioned above.

Let us look more closely.[58] The above angle ACB is known in astronomy as PARALLAX. It is none other than, as has been just shown, the difference and distinction between two observations from two points or positions. Skilled astronomers study, in particular, how far away the Moon may be from the Earth in this manner. Indeed, the calculation is made the same way as above. In the accompanying figure, A is the center of the Earth, as well as an original point

57 Rule of three, known in Latin as *regula trium numerorum, regula proportionum, regula aurea*. See Clavius (1583, 99–108). The *Regula Detri* used by Maestlin in the first example in order to calculate the lengths of AC and BC corresponds to the law of sines, commonly used to solve non-right triangles.

58 In the margin, in red: What is Parallax.

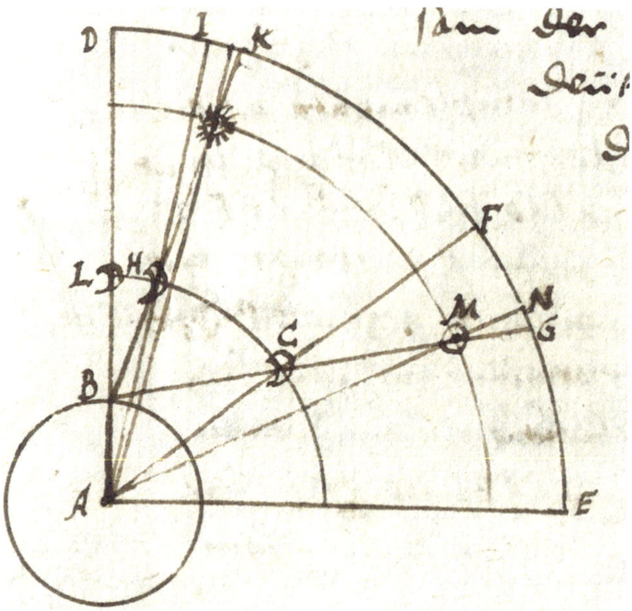

Absehen. B bedeut ein gewißes Ort auf der Erden, als das andere Absehen. C ist ein gewiße Stell des Mons in seiner *Sphaera*. Die lini ACF zaigt im obern himmel DE *verum locum Lunae* underm F. Aber der Mon C wirt aus dem B (gleichsam
21 aus dem andern Stand) gesehen under | dem G. Ist derwegen *Parallaxis* der *Angulus* ACB, oder im obern himmel *arcus* FG. Durch dise *parallaxin* wirt die lini AC, wie weit der Mon von[a] der Erden, oder deren *centro* steht, abgemessen. Dises *calculi* Exempel findt man bey Ptolemaeo *lib*. 5. *cap*. 13. *Almagesti*, und bey *Copernico lib*. 4. cap. 16.

Doch ist hie zu mercken, das ob es wol mit der Cometen *parallaxi* zu observieren, ettwas anderst beschaffen, dann bey dem Mon kan man beede *angulos*, DAC und DBC haben, jenen *per verum motum*, disen *ex observatione*: bey den Cometen aber kan man allein *angulum* DBC observieren. Jedoch was da abgeht, wirt in ander weg erstattet. Nämlich was dorten aus A und B, als aus zweien Ständen wirt gerechnet, das wirt da durch zway *loca* des Cometen, als auch aus zwaien Ständen verrichtet. Dann die *parallaxes* haben die aigenschafft, je weitter sie vom zenith seind, je größer werden sie; je näher je kleiner. Als *angulus*

[a] *In the ms.*: vom.

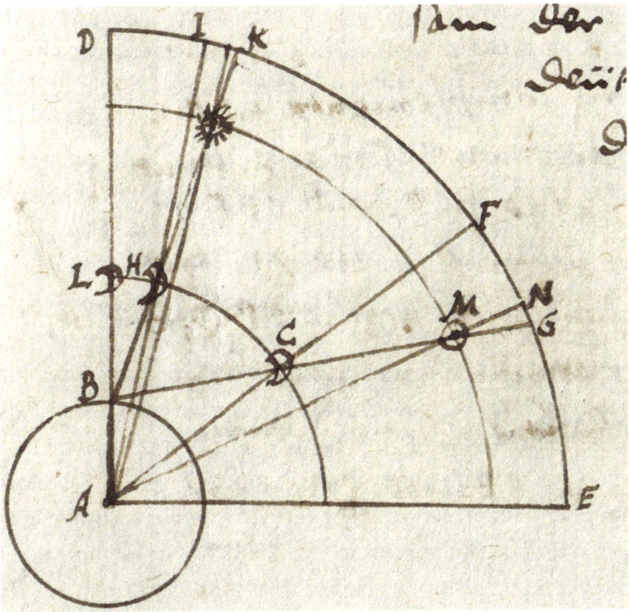

or position. B stands for a certain place on the surface of the Earth, as a second position. C is a certain station of the Moon in its sphere [of motion]. The line ACF shows in the outer heaven DE the actual location of the Moon according to F. Yet the Moon shall be seen from B (as if from the second point) according to | 21
G. The parallax, therefore, is the angle ACB, or the arc FG in the outer heaven. Through [this measure of] parallax, we may determine the length of line AC, or how far away the Moon is from the center or surface of the Earth. You may find an example of this calculation in Ptolemy's *Almagest*, Book 5, Chap. 13, and in Copernicus, Book 4, Chap. 16.[59]

Here we must note, however, that even though parallax should certainly be observed in comets, their case is not completely the same: with the Moon, we can draw both angles DAC and DBC, the former from the actual motion [of the Moon], the latter from observation; with comets, on the other hand, we can only observe angle DBC. Yet the result shall be rendered another way. That is to say, what we may reckon there from A and B, as if from two points, shall be drawn here through two positions of the comet, as if from two points as well. For parallax has the property of growing greater the farther an object is away

59 Cf. Ptolemy (1984, 247–251); Copernicus (1543, 118^{r}–119^{r}); Copernicus (1992, 203–204). The

parallaxeos, in erstgesetztem *schemate*, ACB, sampt sein *arcu* FG, ist grösser als AHB, und sein *arcus* IK. Dann *in linea verticali*, oder im zenith verliert sich die *parallaxis* gar. Als so der Mon ins L kompt, werden AL und BL ain lini, fallen beede *in verticalem lineam* ABD. Nun lehret *Geometria* aus dem underschaid oder dem Exceß *anguli* ACB über AHB, oder *arcus* FG uber IK, die *veram parallaxin* FG, und IK, rechnen, hiemit machet sie auch kundt, wie weit C oder H vom *centro* A stehe.

Darnach ist auch zu mercken, das die *parallaxes* zu observieren nit müglich, es habe dann die distants beeder Absehen A und B, ein *sensibilem proportionem* gegen AC, so zu observieren ist. Dann je weitter es von dannen ist, je kleiner der *angulus parallaxeos* wirt. Als : An dem Mon sind die *parallaxes* vil größer, als an der Sonnen. Dann *angulus* ACB ist größer als AMB, wie auch FG größer ist als NG. Daraus wird bewisen, das der obern *Planeten* höhe durch *parallaxin* nit kan gemeßen werden, dieweil AB die halb Erden Dicke gegen ihrer großen Distants nit zu spiren ist, sonder beede linien AM, und BM, lauffen so nahe zusamen, das zwischen inen zwar ein *angulus* ist, aber selbiger so klein, das er für gar nichts zu halten ist. Ja auch an der Sonnen findt sich die größte *parallaxis* nur 3. scrup. Aber selbige ist *à priore* zu observieren vil zu klein, drum helt mans auch für nichts.

22 Also hin wider, wann die *parallaxis* zu observieren zu klein ist, so ists ein gewiße demonstration und beweisung, das des *corporis* oder Sterns, so observiert solte werden, Distants von beeden Absehen A und B, so groß ist, das gegen ir das AB kein *sensibilem proportionem* hatt.

Dis sind die *Carmina* oder *Incantationes* (wie die poeten fabuliern) damit der Mon und anders gestirn vom himmel herunder gezogen werden, und das seind die Laytern,[a] daran die *Astronomi* gehn himel steigen, und der Sternen[b] höhe abmeßen. Auf dise weis muß man auch mit den Cometen handlen, will man anderst, ob sie *in Elementari oder Aetherea Regione*, hierunden under dem

[a] *Margin*: und ... Laytern. Added at a later date. [b] *Added in the margin:* Sternen.

rough copy (p. 17) elaborates on the second observation (made on 7 August 1524) described by Copernicus on p. 119^{r} and shows that the parallactic angle ACB was 1°. This calculation has been omitted in the clean copy.

60 The manuscript shows an initial *nit* deleted and later added in the margin, near the one deleted as the first word in the line. Maestlin means to say that parallax cannot be observed unless AB maintains an appreciable proportion with respect to AC. The rough copy presents a first draft of this important section on pp. 16–19.

61 Following Aristarchus' and Hipparchus' estimates of the relative sizes and distances of the Earth, Moon and Sun, Ptolemy calculated for the Sun an excessively high parallax of 3 minutes; cf. Ptolemy (1984, 258–265). This value was still accepted by Copernicus and

from the zenith; the closer it is, the smaller the parallax becomes. As we may see in the above diagram, the angle of parallax ACB, along with its arc FG, is greater than AHB and its arc IK. When it is on a vertical line or the zenith, the parallax is lost entirely. Thus, when the Moon arrives at L, AL and BL will form a line, with both falling on the vertical line ABD. Now, geometry shows that, by calculating the difference or excess of angle ACB over AHB, or the arc FG over IK (the actual parallax being FG and IK), we may also hereby show how far away C or H may be from center A.

Next, we must also note that we may not[60] be able to observe parallax unless the distance between points A and B should possess a sensible proportion with respect to AC and may thus be observed. For the farther it is away from them, the smaller the angle of parallax becomes. Thus, parallax is far greater for the Moon than for the Sun. For the angle ACB is greater than AMB, in the same way as FG is greater than NG. Accordingly, we may prove that the distance of the superior planets cannot be measured by parallax, since AB, half the thickness of the Earth, is imperceptible in relation to their great distance, but both lines AM and BM extend so closely to each other that between them there is, in fact, an angle, but it is so small that we may consider it nothing at all. Even in the case of the Sun, the greatest parallax that can be found is only $3'$.[61] However, that is presumably far too small to observe, which is why one may consider it nothing as well.

On the other hand, we may say that when parallax is too small to observe, it 22
is a certain demonstration and proof that the distance of the body or star, as it should be observed, is so great from both points A and B that AB has no sensible proportion in relation to it.

These are the songs and spells (as the poets say) by which the Moon and other stars are drawn down from the sky, and these are the ladders[62] by which the astronomers ascend to the heavens[63] and measure the heights of the stars. We must also deal with comets in the same way if we wish to be certain, whether they dwell in the elementary or the ethereal region, down here below

Brahe; cf. *De revolutionibus*, IV, 21 in Copernicus (1543, 123^r), Copernicus (1992, 208) and Brahe (1602, 65). The true value is $8''$.

62 In true Melanchthonian spirit, Maestlin indicated that Mathematics (Arithmetics and Geometry) is the *Leiter* or wings enabling man to ascend to the heavens. See Melanchthon (1999, 93, 99), Preface to Arithmetic (read by Georg Joachim Rheticus in 1536), and his preface to Johannes Vögelin's *Book on the Elements of Geometry* (1536). See p. 40 below for another mention of this significant topic.

63 Cf. p. 40 below the reference to Arithmetic and Geometry as the two wings of the soul enabling man to fly to the heavens and measure the distances of phenomena from on high.

Mon innerhalb der Elementen, oder droben im himmel versieren, gewiß sein will. Dann ist es, das sie *in regione Elementari* seind, so muß gewiß folgen, das sie größern *parallaxibus* underworfen seyen, als der Mon. Seind sie aber droben, so künden die *parallaxes* so gros nit sein, als des Mons. Were es aber, das man an inen gar kein *parallaxin*, oder *parallaxeos differentiam* spüret, so kans nit fählen, sie seind so hoch das die halbe Erden Dicke gegen selbiger höhe kein *sensibilem proportionem* hatt.

the Moon among the elements or up above in the heavens. For if they are in the elementary region, so must it surely follow that they are subject to a greater degree of parallax than the Moon. If they are up above, however, their measure of parallax could not be as great as that of the Moon. Should we discern in them no difference in parallax or entirely none at all, then it must be the case that comets are so far away that half the Earth's thickness bears no sensible proportion to their distance.

CAPUT VI

Ob diser unser Comet einiche *parallaxin sensibilem* ghabt habe, oder nit: und dann wie weit er von der Erden gestanden möchte sein

Nach Anlaittung fürgebrachten Berichts, und gefundnem rechtem Grund, ist ietzt, alle zweifelhaffte Muttmaßungen hindan gesetzt, nit allein nit unmüglich, sonder aus den *observationibus* gewiß zu erkundigen, ob diser unser Comer under oder ob des Mons *Sphaera* oder himmel sein stell und *sedem* ghabt habe. Hierzu dienet in sonderhait folgende observation.

Den 12. *Decembris* vorigen 1618. Jars, ungefahr umb 10. Uhr folgender nacht,
23 nach dem der Comet underm | *polo* über dem *Meridianum* gegen Orient kommen, und sich im wider auffsteigen aus den düfftigen *vaporibus* herfürgethon, stund er gar nahendt *in recta linea* von vordersten Roß des großen wagens, gegen dem einen sternlin ins *Bootis* lincken hand (num. 2) ohn, das er selbige lini noch nit gar erraicht hatt. Sein *locus apparens, secundum Regulas Astronomicas* gegen disem und andern Sternen, wie sie nach *Tychonis* verzaichnus jetziger Zeit stehn, mit fleis observiert, war 24. grad. 24. scr. ♍, *Latitudo* 57. gr. 13. scr. Folgenden morgen den 13 *Decemb.* umb 5. Uhr, war er fortgeruckt, und stund *praecisè in recta linea* zwischen gemeldtem Roß und dem ersten Stern[a] (num. 1.) ins *Bootis* lincken Hand. Sein *locus apparens* gleicher maßen gegen disen und andern Sternen observiert, war 23. gr. 26. scr. ♍, *Latitudo* 57. gr. 35. scr. Ist also der Comet in 7. stunden in seinem gang fortgeschritten *secundum longitudinem* gar nahendt ain grad (58. scr.) und *secundum latitudinem* 22. scr. welcher *motus* erstreckt sich in seinem *motus proprii circulo,* uff 38 scr. Eben so vil, nit weniger noch mehr, hatt auch sein *motus proprius* ertragen. Dann vorigen morgen, den 12. *Decemb.* war er umb 5. uhr in 26. gr. 44. scr. ♍, *cum latitudine* 56. gr. 19 scr. Hieraus ist gewiß und unfählbar zuschließen, das in disen 7. stunden, von 10. uhr zu Abendt, bis 5 uhren am Morgen, sich gantz und gar kein *parallaxeon differentia* an disem Cometen erzaigt hatt. Dann am Abendt umb 10. uhr, war die Sonn im 1. gr. 8. scr. ♑, deshalben nach ausweisung *Calculi Astronomici,* aus den *Tabulis primi motus,* und *Triangulorum doctrina,* war

[a] *Margin*: Stern.

64 Cf. p. 13 above and notes 36, 37, as well as Figure 5.

65 Brahe (1602, 268).

66 Taking $\varphi = 48° \ 31'$, $\lambda = 9° \ 3' \ 21''$ E = − 0 h 36 m 13 s as modern geographic coordinates for Tübingen, where Maestlin observes, 22 h at Tübingen correspond to 21h 24m UT (22 December 1618 old style). At this moment, the modern coordinates of the comet are

Chapter 6
Whether Our Present Comet Possessed Any Sensible Parallax or Not, and How Far Away It May Have Been from the Earth

Following the course of the foregoing account and the proper foundation for it, it is now time to leave behind any doubtful conjectures and, based on the observations, inquire not only possibly but with certainty whether our present comet occupied a place and position below the sphere of the Moon or above it in the heavens. Here, the following observation is particularly important.

On 12 December of the previous year (1618), around 10:00 the following night, after the comet had passed beneath | the pole above the meridian toward the 23
East and begun to rise again from the moist vapors, it stood almost exactly in a straight line from the foremost steed of the Big Wagon, alongside a little star in the left hand of Boötes (no. 2),[64] without quite yet reaching the straight line. The apparent location of the comet, observed diligently according to the astronomical rules [*regulae astronomicae*] regarding this and other stars as they are currently situated according to Tycho's catalogue,[65] was 24°24′ Virgo, with a latitude of 57°13′.[66] The following morning, at 5:00 on 13 December, it had moved further and stood precisely on a straight line between the aforesaid steed and the first star in the left hand of Boötes (no. 1).[67] The apparent location of the comet, observed in the same way according to this and other stars, was 23°26′ Virgo, with a latitude of 57°35′. Thus, in seven hours the comet had proceeded on its path by almost an entire degree in longitude (58′) and 22′ in latitude, a measure of motion that covers 38′ on its circle of proper motion. This is the rate at which the proper motion of the comet remained, neither more nor less. For the previous morning, at 5:00 on 12 December, it was at 26°44′ Virgo, with a latitude of 56°19′. We may therefore conclude clearly and unequivocally that in those seven hours, from 10:00 in the evening until 5:00 in the morning, there occurred absolutely no difference in parallax in this comet. For at 10:00 in the evening the Sun was at 1°8′ Capricorn, so according to the demonstration of astronomical calculation, following the tables of the first motion, and

longitude 23° 50′ Virgo, latitude 57° 09′. That day, towards 22 hours at Tübingen, the Moon sets on the western horizon, so that its zenith distance amounts to nearly 90°. The first quarter fell on the 24 December, so that the Moon was close to quadrature, explaining the value of the minimum parallax sought.

67 Ptolemy (1984, 346), ϰ Boo in modern designation.

Ascensio Recta Solis 271. temp. 13′. *distantia à Meridiano* 150. temp. Unde *Asc: Rec: Medii Coeli* 61. temp. 13′. Also *locus Cometae* 24. gr. 24′. ♍, *Latit.* 57. gr. 13′, gibt sein *declinationem* 52. gr. 21′, *Asc. rect.* 208. temp. 9′. und deswegen sein *dis-*
24 *tantiam* gegen Orient 146. temp. 56′. aber vom Ze|nith 75. gr. 18′. Also folgenden morgen umb 5. uhr, war *locus Solis* 1. gr. 26′. ♑, *Asc. Rect.* 271. temp. 33′. *distantia Solis à Meridiano in consequentia* 255. temp. *Asc. R. M. C.* 161. temp. 33′. Des Cometen *locus* 23. gr. 26′. ♍ cum *latit.* 57. gr. 35′., gibt sein *declinationem* 52. gr. 59′. *Asc. R.* 207 temp. 13′. und *distantiam à Meridiano* 40 temp. 40′. et à Zenith 25. gr. 46′. Wann nun der Comet were *in regione elementari* gestanden, oder auch zu höchst derselben zu aller nächst under dem Mon (welche *distantiam Copernicus lib.* 4 *cap.* 17 erweiset, das sie seye 52. *semidiam. terrae,* 17 scr.) so were in der ersten observation, umb 10 uhr, sein *parallaxis,* weniger nicht, als *Lunae proximae,* wie *Tab. Prut.* ausweisen 1. gr. 5′. In der andern aber 0 gr. 29′. Were also *parallaxeon differentia* von 10. uhr, bis 5. uhr gweßt 36. scr. Hette derwegen die *parallaxis* in disen 7. stunden sich umb 36. scr. verlohren, und hette *Cometae motus apparens* nicht nur 38. scr. sonder 1 gr. 14′. sein müßen, welche so große Differentz were freilich wol zu observieren geweßt, und unmüglich, das in fleißiger observation ein so großer und gleichsam handgreiffiger underschaid hette könden übersehen werden. Solte dann der Comet noch näher bey der Erden gestanden sein, so müßte ja die *parallaxis* und ire Differentz auch desto größer, und desto beßer zu observieren gweßt sein. Aber des Dings hatt sich gar nichts erzaigt, sonder *motus verus et apparens* haben umb kein minuten differiert.

Gleicher gestalt hatt sich diser Comet auch zu andern Zeitten erzaigt, sonderlich den 3. *Decemb.* als er bey den zweien sternlin auff des *Bootis* Rucken (num. 17. und 18.) und den 19. *Decemb.* bey den zweyen sternlin uf des Drachen

---

68 The modern value for the longitude of the Sun at 5h for Tübingen (= 4h 24m UT) is 1° 24′ Capricorn (that is, 271° 24′). The *Prutenic Tables* (which locate Tübingen 1h 03m of western longitude from the meridian of reference: when it is 5h at Tübingen, it is then 6h 03m at Königsberg) assign to the Sun a longitude of 270° 49′. The deviation from the *Prutenic Tables* is important, for it may suggest that Maestlin used other tables for establishing the location of the Sun.

69 Modern values: longitude = 23° 01′ Virgo and a latitude of 57° 27′.

70 Copernicus (1543, 120^{r}), Copernicus (1992, 205).

71 Reinhold (1551, 137^{r}); the same table in *De revolutionibus*, IV, 24, with slightly different values; see Copernicus (1992, 213). The exact figure, according to Reinhold's table for the parallax of the Sun and Moon, is 1° 6′ 21″ = 1° 6′. However, Maestlin rounds to 1° 5′. The modern value for the parallax of the Moon in Tübingen on 22 December, at 22 hours, is

the doctrine of triangles, the right ascension of the Sun was 271°13′ and its distance from the meridian was 150°. Thus, the right ascension of the midheaven [*medium coeli*] was 61°13′. Accordingly, the location of the comet, at 24°44′ Virgo, with a latitude of 57°13′, yields a declination of 52°21′ and a right ascension of 208°9′, and thus the distance of the comet was 146°56′ toward the East, 75°18′ from the zenith |. Then at 5:00 the following morning, the Sun 24
was located at 1°26′ Capricorn,[68] with a right ascension of 271°33′; the distance of the Sun from the meridian was 255° according to the order of the signs and the right ascension of the midheaven was 161°33′. The location of the comet, at 23°26′ Virgo, with a latitude of 57°35′,[69] yields a declination of 52°59′, a right ascension of 207°13′, and a distance from the meridian of 40°40′ and from the zenith of 25°46′. Now, if the comet had been situated in the elementary region, or even in the most remote part of it just beneath the Moon (the distance of which Copernicus demonstrates to be 52 earth radii and 17′ in Book 4, Chap. 17),[70] so would its parallax in the first observation, at 10:00 [in the evening], have been no less than the nearest measure of the Moon, as the *Prutenic Tables* prove to be 1°5′.[71] In the second [observation], it would have been 0°29′.

Consequently, the difference in parallax from 10:00 [in the evening] to 5:00 [in the morning] had been 36′. The parallax over those seven hours would therefore have decreased by 36′ and the apparent motion of the comet should have been not only 38′ but 1°14′, a difference so great that it would, in fact, have been plainly observed and not been possible for such a great and discernible difference to have escaped close scrutiny. Should the comet have been situated even closer to the Earth, then the parallax and its difference would have been even greater and all the easier to observe. Yet no such thing occurred, and the true and apparent motion [of the comet] did not differ by a minute.

The comet proceeded in the same way at other times, particularly on 3 December, when it drew near the two small stars in the back of Boötes (nos. 17 and 18), and on 19 December, when it approached the two little stars in the tail

---

0° 56′. With the *Prutenic Tables* we arrive at a parallax of 1° 0′ 45″, around 4′ higher, an acceptable difference. The parallax of 1° 0′ 45″ has been calculated following Reinhold's accurate method: the theory of the Moon in the *Prutenic Tables* has been used, which assigns to Tübingen on 22 December 1618 at 22 h (that is, 23h 03 m in Königsberg, the meridian of reference in the *Tables*) a mean elongation of 72° 47′, an anomaly of 109° 50′ and a zenith distance of 90° for the Moon. This does not explain why Maestlin rounds 1° 5′ 45″ down to 1° 5′. This parallax corresponds to an Earth-Moon distance of 52/53 e.r. We are grateful to Denis Savoie for his advice on this and the previous notes.

schwantz (num. 28. und 29.) wie droben[a] gemeldet worden, ist hingezogen, daselbsten waren auch zu underschidlichen stunden der nacht *motus verus et apparens* einander aller dings gleich gefunden.

Aus disem allem ist unwidersprechlich zu schließen, das diser Comet von der Erden *in aethercam regionem* |[b] so hoch erhebt gweßt sey, da sich alle *parallaxeon sensibilitas* verliert, und deswegen von und ob der Erden ferner und höher gestanden, als die Sonn von der Erden ist, das ist, Er hatt nicht weniger als 1200. *semidiametris Terrae*[c] (welche, jeden *semidiametrum terrae* zu 860. teutscher meilen gerechnet, thun mehr als tausent mal tausent (1.000.000) teutscher meilen) von der Erden seinen Lauff gehabt.

Nach dem nun also erwisen ist, das diser Comet mit nichten *in Sphaera elementari*, sonder seer hoch droben im Himel, gestanden seye (ob er aber, und umb wie vil er höher, als die Sonn seye gweßt, ist nit zuerforschen, dann die *parallaxes* lauffen allhie aus) so ist jetzt die nächst frag bey meniglich,[d] wie lang raichen seine Strahlen? wie weit hatt Er seinen Bart oder Haarlöck von sich ausgestrecket? Antwort, aus gewißer observation. Den 27. *Nouembr.* frue umb 5. uhr, war des Cometen Haupt oder Stern (wie auch droben angezaigt)[e] im 29. gr. 26 scr. ♎, *habens latitud.* 28. gr. 50′. Dazumal erstrecket er sich bis hinan, und erraichet den Stern auf des großen Beeren hufften (num. 22.). Es ist aber wol gläublich, er seye noch weitter hinaus gangen, dann dieweil er zu end gar dünn und dunckel geleuchtet, hatt man sein eusserst end, sonderlich wegen seiner überaus großen höhe, nit gwis zaichnen könden. Es steht aber diser Stern im 22. gr. 48′. ♌, *habens latitudinem* 35. gr. 14′. Hieraus gibt *doctrina triangulorum* zu erkennen, das der Cometstern, oder sein Haupt ist von disem Stern gestanden gantze 56. grad. *circuli in coelo magni.* War also dises Cometen lenge nicht weniger als 56. grad. Sein braite erraichet auch über 6. grad. Jetzt gesetzt, der Comet were nit ferner von der Erden, als nur 1200. *semidiametris terrae* geweßt,

---

[a] *Margin*: *Cap.* 3. [b] Here the fourth sheet in the clean copy begins. [c] *Margin*: *Cometae distantia* von der Erden. [d] *Margin*: Wie lang der Comet sich ausgestreckt hab. [e] *Margin*: *Cap.* 3.

---

72 *Margin*: *Cap.* 3. Cf. p. 13 above.

73 *Margin*: Distance of the comet from the Earth. This addition as well as the following ones were also made at a later date. 1210 earth radii was the distance of the Sun according to Ptolemy (1984, 257). Copernicus (1543, 122[r]; 1992, 207) had calculated 1179 e.r. for the distance to the Sun at apogee and Maestlin, in his Appendix "On the dimensions of the celestial orbs and spheres" to Kepler's *Mysterium cosmographicum* [Kepler (1596, 136)] had calculated 1208 e.r. See Van Helden (1985, 46, 48, 60, 72–73).

74 Chapter 6 up to this point coincides with pp. 18–20 in the rough copy. The rest of the chapter is a new development. For the issue of the parallax of the comet, see also Schickard (1619, 160–162), who grants the comet a parallax of a few minutes and locates it between

of the Dragon (nos. 28 and 29), as mentioned above;[72] on those days, the true and apparent motion was found to follow precisely the same pace at different hours of the night.

We may conclude irrefutably from the above analysis that the comet was so far removed from the Earth in the ethereal region | that any hint of parallax was lost entirely, and therefore that it was situated far above and beyond the Earth than the Sun is from the Earth, that is, the comet followed a course no less than 1,200 earth radii from the Earth[73] (which, if we reckon every earth radius as 860 German miles, makes more than 1,000,000 German miles).[74]

Now that is has been proven that the comet was not even near the elementary sphere, but very high above in the heavens (whether and how much higher it was than the Sun cannot be determined, however, for here the parallax peters out), the next question now by far is how long its rays of light extended, or how far it stretched out its beard or locks of hair. Answer, from certain observation. At 5:00 in the morning on 27 November, the head or star of the comet (as also shown above)[75] was at 29°26′ Libra, with a latitude of 28°50′. At that time, it stretched all the way back to reach the star in the hip of the Great Bear (no. 22). It is quite plausible, however, that it extended even further, for since it then shone so dark and thin at the end no one could have determined its full extent with certainty, especially given its extremely great distance. In any case, the above star stood at 22°48′ Leo, with a latitude of 35°14′.[76] According to the doctrine of triangles, we may thus establish that the head or star of the comet was situated fully 56° away from this star on a great circle in the heavens. That is to say, the length of the comet was no less than 56°. It spanned more than 6° in width as well. Now, assuming that the comet was not farther away from the

---

Venus and the Sun: "der Comet schier keine, oder doch gar ein kleine und geringe *parallaxin* müsse gehabt haben. [...] er hab so gar ein geringer *parallaxin*, nur ettlich wenig *minutorum* ghabt. Weil die *observationes*, beedes [p. 161] *prope horizontem*, und *longe supra illum* geschehen, (da es ja solt in *sublunaribus* ein *maximam evariationem visus* gegeben haben) und aber alle solche sechs Ort, beynahend in eine gerade Lini fallen (welches im Himmel droben ein *circulus maximus* ist) so versteth sichs ja hierauß, es hab die Erddikhin gegen des Cometen mächtiger Höhin ein gar schlechtes, oder schier ein unmerkhlichs außgetragen, und sey er gar weit ob dem Mond, in *ipso aethere* gestanden. Dan were er *sublunaris* gwest, so hetten solche *loca* freylich nimmermehr in eine Lini zusamen treffen könden, sonder weren schier ein Grad hin und wider exorbitirt. Doch muoß ich ihne auch nit gar zu hoch hinauff erheben, und will deßwegen *aliâ et noviter excogitatâ ratione* hirmit beweisen, daß er zwar wie gemelt *supra Lunam*, aber doch *infra Solem*, gewest sey. [...] der Comet, wie bißher erwisen, zwar under der Sonnen, doch weit ob dem Mond, aber nit gerad in der Mitten darzwischen gestanden, sonder ettwas höhers".

75 *Margin*: *Cap*. 3. See pp. 13, 14 above.

76 Cf. p. 14, note 44 above.

so gibt *Regula Detri*: wie sich helt ain *semidiameter terrae*, gegen 15 teutscher meil (das ist gegen einem grad *in circulo magno* auf der Erden, da ain grad 15 meil helt) also halten sich 1200. *semidiam.* gegen ainem grad ins Cometen
26 *circulo* oder *Sphaera*. | Nach ausweisung der Regel, kommen jedem grad achtzehen tausent (18000.) meilen. Dise in 56. grad multipliriert, geben die lenge des Cometen mehr als tausent mal tausent teutscher meilen, und die braite mehr als hundert mal tausent meilen. Sihe ein solches mächtig und erschrocklich liecht oder fewr, welches doch in kürtze wider verleschen solt, kan und hatt der Allmächtige Gott am Himmel angezündet. Warumb? One zweifel seinen grimmigen zorn uber unsere Sünden uns Menschen anzuzaigen. Wer wolte dann nit erzittern und sich fürchten, wan ein solches fewr angehet, und diser Lewe zu brüllen anfehet? Aber hiervon werden wir aus Gottes wort sonsten gnug erinnert, laßt uns solches zu Hertzen gehen, und uns zu Ihm bekheren.

Hie kan ich mich nit gnug verwundern, wa doch ettliche unserer jetzigen *Mathematicorum* ire gesicht und *iudicia* hingethon haben, welche in irem observieren diesen Cometen nit lenger befunden oder geschetzt haben als 20. oder 30. 32. 39. oder aufs höchst 40. grad lang. Da er doch vor und nach dem 27. *Nouemb.* eben dise lenge auf 56. grad erraicht hatt, und sonderlich an selbigem tag den obgenanten Stern auffs großen Beeren hüfften augenscheinlich angetroffen.

Earth than only 1,200 earth radii, the rule of three grants that, in the same way an earth radius stands in relation to 15 German miles (that is, in relation to 1° on the great circle around the Earth, since 1° is equal to 15 miles), 1,200 earth radii may stand in relation to 1° on the circle or sphere of the comet. | According to 26
the demonstration of the rule, each degree amounts to 18,000 miles. Multiplied by 56°, the length of the comet measures more than 1,000,000 German miles and the width more than 100,000 miles. Behold such a mighty and terrifying light or fire, which should soon go out again, Almighty God can and has kindled in the heavens. Why? No doubt to express to us men his fierce wrath over our sins. Who then would not wish to quake and tremble with fear when such a fire is set and this lion begins to roar? Yet by it we are duly recalled to the word of God, let it enter our heart and convert us to Him.[77]

Here, I cannot cease to wonder what several of our mathematicians today have set their sight on and subjected to judgement, finding or estimating in their observation of the comet nothing longer than 20° or 30°, 32°, 39°, or at the most 40° in length. For it reached the very same length of 56° before and after 27 November, and specifically on the same day it apparently encountered the above mentioned star in the hip of the Great Bear.

77 See p. 1 above (the first page in the rough copy), with the reference to comets as signs (*Zeichen*) of God's wrath against the sins of mankind and as stimuli for repentance.

CAPUT VII

Das auch vor diser Zeit vil andere Cometen nicht im Lufft, sonder im hohen Himmel erschinen, und observiert worden seind

Ab disem was erzehlt ist worden, verwundern sich ettliche nit wenig, wie doch ein Comet im hohen himmel generiert werde? und wie die *vapores, fumi*, oder *exhalationes*, die dämpf so von der Erden auff fahren, könden so hoch hinauf steigen? Aber ehe von den exhalationen, düfften oder dämpfen weitter ettwas gemeldet werde (dann es ist noch nit erwisen, das die Cometen aus den *exhalationibus*, so von der Erden übersich riechen, iren uhrsprung, substantz und wesen haben, welches doch sie gleichsam für gwiß und bekandt halten) so ist
27 hie zuvor | zu mercken, das dises nichts news ist, sonder wie sich diser unser Comet im jar 1618. erzaigt, also finden sich seins gleichen noch mehr andere, an welchen eben so wenig zu zweifflen ist, das sie nit im lufft, sonder in *aetherea regione*, im himel gestanden seien.

Erstlich,[a] ists gewiß, das seid von *Anno* 1572. alle Cometen und newe Stern, nämlich die Cometen *Anni* 1577. 1580. 1582. 1585. 1590. 1596. 1607. sampt den Sternen 1572. und 1604, sind in *aethere*, und mit nichten in *regione elementari* geweßt. Dann bey disen allen ist kein *parallaxis* gespirt worden. Hieher gehört auch der Comet *Anni* 1556.[b] Das auch diser on alle *parallaxi* gweßt sey, kan aus fürnemer *Mathematicorum*, sonderlich *Pauli Fabricii*, und *Joachimi* Hellers *observationibus* erwisen werden, deren jener zu Wien, diser zu Nürnberg, bisweilen in einer nacht, zu underschidlichen stunden, da der Comet das eine mal gegen Aufgang, das ander mal über den *Meridianum* gegen Nidergang gestanden, ine observiert haben. Da dann wan einige *parallaxis* were zu spiren gweßt, sie sich würde endeckt haben. Aber hiervon soll in einem andern Tractat, ob Gott will, gehandlet werden.

Von den Cometen,[c] so in *annis* 1531. 1532. 1533. 1538. 1539. geschinen, bezeugts eben dis Petrus Apianus, dazumal Kaysers *Caroli* V *Mathematicus*, welcher sie observiert hatt. Es war zwar *Petrus Apianus* anfänglich auch diser meinung, die Cometen werden nirgendts anderstwo als im Lufft, oder *sphaera elemen-*

---

[a] *Margin*: *1.* Alle Cometen von Anno 1572. sind nit im lufft, sonder in aethere gestanden. From here up to the mention of Andreas Nolthius on p. 29, the text coincides with pp. 21–22 in the rough copy. [b] *Margin*: also der Comet Anno 1556. [c] *Margin*: 2. Cometen von 1532. bis 1539.

---

78 *Margin*: *1.* All comets from the year 1572. have not been in the air, but in the ether. From here up to the mention of Andreas Nolthius on p. 29, the text coincides with pp. 21–22 in the rough copy.

79 *Margin*: also the Comet of 1556.

Chapter 7
That before This Time Many Other Comets Appeared and Were Observed Not in the Air, but in the Upper Heaven

From what has been said above, many wonder more than a little how a comet is generated in the upper heaven, and how the exhalations, fumes, or vapors that come rising damply from the earth could ascend so high. Yet before further mention is made of the exhalations, fumes, or vapors (for it has not yet been proven that comets draw their origin, substance, and being from the exhalations that emanate from the earth, even though they are held as such for certain and well established), we should first | note that this is nothing new; but in the 27
same way our present comet of 1618 arose, so many others may be found of this sort in which there is just as little doubt that they were situated not in the air, but in the ethereal region in the heavens.

First of all,[78] it is certain that since the year 1572 all comets and new stars, namely the comets of 1577, 1580, 1582, 1585, 1590, 1596, and 1607, along with the stars of 1572 and 1604, have been in the ether, and by no means in the elementary region. For there was no sign of parallax in any of these. The comet of 1556 belongs among them as well.[79] That the latter did not possess any parallax can be proven from the observations of distinguished mathematicians, particularly Paul Fabricius and Joachim Heller, the former in Vienna, the latter in Nuremberg, who observed it at different times over the course of one night, when the comet was situated at one point toward the East and at another point above the meridian toward the West.[80] For if any parallax would have been there to discern, it would have been discovered. But this should be dealt with in another treatise, God willing.

As for the comets[81] that appeared in the years 1531, 1532, 1533,[82] 1538, and 1539, we may find testimony from Petrus Apianus, at that time mathematician to [Holy Roman] Emperor Charles V, who observed them.[83] At first, Petrus Apianus also held the opinion that comets are kindled in no other place than

80 On this comet, see Hellmann (1944, 106–111), where the observations of both Fabricius and Heller are mentioned. On Fabricius, see also Jerratsch (2020, 155–162).

81 *Margin*: 2. Comets from 1532 to 1539.

82 Maestlin has corrected a prior 1535. On p. 21 in the rough copy, *1533*.

83 For a general study on Petrus Apianus, see Röttel (1995). On his works on comets, see Kokott (1994, 56–72); Hellmann (1944, 88–91); Barker (1993, 6–10), Jerratsch (2020, 122–127).

tari entzündt. Aber in betrachtung Ihrer lauffs und *parallaxeos*, hatt er bald angefangen zu zweiflen. Dann[a] da er gemeint, die Cometen solten, mit des fewrs, und obere thails des Luffts *sphaeris* (als vons Mons *sphaera* getriben) *secundum signorum ordinem* iren aignen lauf haben; So hat der Comet *Anni* 1533. das widerige gewisen, dan Er ist *contra signorum ordinem*, von Zwiling in Stier gangen. Darumb (wie Er in *Caesar. Astron.* fol. O pag. a bekhennet) hat er schon dazumal anfangen zu zweiflen, ob die *Sphaera Ignis*, und oberer Lufft vom Mon getriben werden? Und wolte schier nit ungern disen *philosophis* beyfallen, welche lehrten, die Cometen rucken von und zu der Sonnen, das ist, sie schweben nit im Lufft, sonder seyen droben im Himmel, nit weit von der Sonnen. Aber nach dem er deßen gnugsam bey sich überzeugt war, ist er[b] gar anders gesinnet worden. Deswegen sein *filius, D. Philippus Apianus, Mathematum* alhie zu Tübingen gewesner Professor, und mein lieber *Praeceptor* seliger gedächtnus, zehlet seinen Vatter auch under die fürneme *Mathematicos*, welche durch gewiße und unfählbare *observationes* überzeugt, haben bekennen
28 müßen, die Cometen | werden im Himmel generiert. Als da sind *Vitus Amerbachius, Gemma Phrisius, Antonius Mizaldus, Johannes Pena, Thaddaeus Hagecius*, etc.

Hie irret nicht,[c] das Johannes Vögelin von Hailbronn, sonsten ein fürnemer *Mathematicus* zu Wien, den Cometen *Anni* 1532. gleichsam mit gwalt von himmel herunder in den Lufft gezogen, das ist, das er in einem besondern Tractatlin

---

[a] Long addition at the bottom of the page, absent from the rough copy (p. 21): Dann [...] nit weit von der Sonnen. [b] *Deleted*: letztlich. [c] *Margin*: Johannes Vögelin hatt sich mit dem Cometen 1532. grob verstoßen.

---

84 Apianus (1540, fol. $O3^{r}$). Apian's cometary observations and conclusions are described in fol. $N2^{v}$– $O3^{r}$.

85 Apianus (1540, fol. $O3^{r}$): "Ex istiusmodi motu [of the 1533 comet] certe non difficile fuerit iudicare, igneam et supremam aëris sphaeram non utique ab impetu Lunaris sphaerae rapi, alioquin enim Cometam ab occasu in ortum ferri par erat".

86 Philip Apianus (1531–1589) had to leave the (Catholic) University of Ingolstadt in 1569 due to his Protestant (Calvinist) convictions. He lectured at Tübingen from 1570 to 1583, where he taught Maestlin, but was forced to renounce his position for refusing to sign the Lutheran Formula of Concord. His renunciation allowed Maestlin, who had been professor of mathematics at the University of Heidelberg since 1580, to join the faculty at Tübingen. Maestlin bought some books belonging to Philip Apianus from his widow, including one in which he found a note attributing the praefatiuncula to Copernicus' *De revolutionibus* to Andreas Osiander. On this, see Gingerich (2002, 220–221).

87 On Veit Amerbach (1503–1557), see Frank (1997) and Burmeister (2015, 68–69). Amerbach was successively professor in Wittenberg and Ingolstadt. Most probably, Maestlin refers to his edition and commentary of Pontanus, *Liber de meteoris*, where, without examining

the air or the elementary sphere. But once he considered their course and parallax, he began to doubt. For Apianus first thought that comets, with the spheres of fire and the upper part of the air (as they are driven by the sphere of the Moon), should follow their own course of motion according to the order of the signs; however, the comet of 1533 proved otherwise, for it proceeded contrary to the order of the signs, from Gemini to Taurus. For this reason (as he confesses on fol. O, page a, in the *Astronomicum Caesareum*),[84] he began to doubt from that time whether the sphere of fire and the upper air were driven by the Moon.[85] And he simply did not wish to follow reluctantly those philosophers who taught that comets turn toward and away from the Sun, that is, that they do not float in the air but are up in the heavens, not far from the Sun. After he was sufficiently convinced of the matter, however, he was disposed quite differently. That is why his son, Philippus Apianus,[86] former professor of mathematics here at Tübingen and my beloved teacher of blessed memory, counted his father also among those leading mathematicians who, convinced by certain and infallible observations, were compelled to admit that comets | are generated in the heavens. 28
Among others, we may count Vitus Amerbachius,[87] Gemma Frisius,[88] Antonius Mizaldus,[89] Johannes Pena,[90] and Thaddaeus Hagecius.[91]

Here, it is true[92] that Johannes Vögelin of Heilbronn, an otherwise distinguished mathematician in Vienna, violently pulled the comet of 1532 down from the heavens to the air, that is to say, in a particular little treatise[93] he per-

---

any particular comet, he expresses his preference for Seneca's doctrine on the celestial character of comets contrary to the Aristotelian concept. See Amerbach (1545, 199–200).

88 Gemma Frisius (1508–1555) adhered to Apianus' optical theory of comets, without, however, determining their celestial or elementary location. See Frisius (1545, 33^{r}–^{v}), Barker (1993, 10–11).

89 Antoine Mizauld (1510–1578) was a conservative, Aristotelian defender of the traditional concept of comets as inflamed exhalations in the upper elementary region. See Mizauld (1549, 42–43). Occasionally, however, he refers to ancient authors (in particular, Seneca) who were supporters of the celestial concept of comets and Regiomontanus on the celestial location of the comet of 1475 [actually 1472]; see Mizaldus (1549, 79–81, 95–98).

90 Jean Pena (1528 or 1531–1558), wrote a famous preface to his edition and translation of Euclid's *Optics* (1557). In his preface *De usu optices* he affirmed that the celestial medium through which planets move was air and that at least some comets were located above the Moon. See Pena (1557, fol. bb ii^{v}), Barker (1993, 11–15), Granada (2002, 6–23).

91 Taddaeus Hagecius ab Hayck (1525–1600) was the foremost astronomer of eastern Europe in the second half of the sixteenth century. In his very significant treatise on the nova of 1572, he argued that comets can be located in both the celestial and elementary regions; see Hagecius (1574, 46–59) and here, p. 29 below. On Hagecius and his relation to Maestlin, see Hellman (1944, 184–206 and *ad indicem*).

92 *Margin*: Johannes Vögelin went grossly astray with the comet of 1532.

93 Cf. Vögelin (1533). On this work, see Hellman (1944, 96–97), Kokott (1994, 98–99, 135–136).

sich selbs, und andere, beredet, und vermeint, er habe gleich wol, künstlich und gewiß demonstriert, selbiger Comet sey vom *centro terrae* nur 1535. teutscher meilen, und seiner rechnung nach ob der Erden nur 640. teutscher meilen gestanden. Aber so man seine *observationes* und sein Rechnen ettwas fleißigers besucht, findt es sich gar bald, das er sich grob gnug verstoßen hatt. Dann erstlich, nimpt er die zeit zwischen seinen zwaien observationen nit gar drey viertel einer stund, welche zeit zu solchem *calculo*, die *differentiam parallaxeon* zu underschaiden, und daraus die *veram parallaxin* zu rechnen, vil zu kurz ist. Darnach sein *modus observandi per solos circulos verticales et altitudinem*, ist gar zu schlüpferig, und mag darin gar leichtlich geirret werden, dann selbiger fordert scharpsinnigen fleiß, also wa darinnen, an der zeit, an der höhe, an gewißer Stell, oder Abzaichnung *lineae Meridianae*, und an der distantz *verticalis circuli à Meridiano*, nur ettlich minuten übersehen werden, so ists bald umb ettlich tausent meil im rechnen gefählt. Aber Er, Vögelin, achtet solcher subtilitet gar wenig, sonder laßt seine observationen nur bey gantzen, oder halben graden bleiben. Wie wolt er dann ettwas gewiß bewisen haben⟨?⟩ Zum dritten, welches noch mehr, sein aigner *calculus* stimmet mit ihm selbs nit überein, wie solchs, wer in examiniert, findet.

Gleicher gestalt[a] haben ettliche sich hefftig bemühet, ob und wie sie doch möchten und köndten dem Cometen *Anni* 1577. ein *locum* hie unden im Lufft einraumen, wie sie aber es getroffen, kan (anders alles zugeschweigen) aus
29 disem | gnugsam abgenommen werden, das sie selbigen Cometen in so ungleiche Losament im Lufft haben einlosiert.

Andreas Nolthius[b] *cap.* 3. nemmet die zeit zwischen seinen zwaien observationen, ain stund, daraus stelt er den Cometen, cap. 5. vom *centro terrae* 8. *semidiametros terrae* und zway drittail, das seind, seiner Rechnung nach 7762. meil, und zway drittail einer meil.

Bartholomaeus Scultetus[c] rechnet zwar mit überaus mühsamer, doch gar unnützer und vergebenlicher Rechnung, die zeit zwischen seinen zwayen observationen 12. minuten, 45 secund. und 28 tertz, einer stund, das were nicht gar ein viertel stund, und bringt den Cometen von *centro terrae* 9. und ein drittail *semidiam. terrae*, welches im sind 8047. meil, und 1754. schritt. Ist aber dis nit gar subtil ausgecirckelt, und doch so übel getroffen?

---

[a] *Margin*: Mit dem Cometen Anni 1577. haben ettliche nit wenig gefählt. [b] *Margin*: Andreas Nolthius. [c] *Margin*: Bartholomaeus Scultetus.

---

94 Vögelin (1533 [in Hagecius 1574], 159 f.).

95 Vögelin (1533 [in Hagecius 1574], 154). The two observations were made on 6 October 1532 at 16h, 8 minutes and at 16h, 50 minutes.

suades himself and others by presuming that he has demonstrated clearly and cleverly that the above comet was only 1,535 German miles away from the center of the Earth and, according to his calculation, just 640 German miles away from the surface.[94] Yet if we examine his observations and calculations more closely, it soon becomes clear that he has gone far enough astray. For, first of all, he does not even take three-quarters of an hour between his two observations,[95] which, for such a calculation, to distinguish the difference of parallax and reckon the real parallax from it, is far too little time. Second, his mode of observation by single vertical circles and altitude[96] is far too slippery and may thus have led quite easily to error, for this demands an acute degree of diligence, since what is subject to scrutiny, including the time, distance, given location, or determination of the meridian line, and distance of the vertical circle from the meridian, has no sooner missed by a few minutes than it errs by thousands of miles in calculation. Yet Vögelin pays very little regard to such subtlety, but allows his observations simply to stand at whole or half degrees. How, then, did he wish to prove anything for certain? Third, and what is even more, his own calculation does not agree with itself, as anyone who examines it may find.

In the same way,[97] many have tried hard, however they could and would, to assign to the comet of 1577 a location down here in the air; but as they determined it, we can (without further mentioning other factors) | deduce suffi- 29
ciently from this that they accommodated the comet in very different dwelling places in the air.

Andreas Nolthius[98] in Chap. 3 takes an hour as the time between his two observations, from which he locates the comet in Chap. 5 8 $^2/_3$ earth radii from the center of the Earth, that is, 7,762 $^2/_3$ miles according to his calculation.

Bartholomaeus Scultetus[99] reckons according to an extremely arduous, but ultimately useless and vain calculation, the time between his two observations as 12 minutes, 45 seconds, and 28 thirds (of an hour), not even a quarter hour, and removes the comet from the center of the Earth by 9 $^1/_3$ earth radii, that is, 8,047 miles and 1,754 feet. But is this not rounded out in no subtle way, and still determined so poorly?

---

96 Vögelin (1533 [in Hagecius 1574], 153 f.).

97 *Margin*: With the comet of 1577, several [observers] have missed not a little.

98 *Margin*: Andreas Nolthius. Cf. Nolthius (1578). On this work, see Hellman (1944, 216–225). Maestlin's criticism was already present in Maestlin (1578, 14–16). For a commentary, see Hellman (1944, 150 f.).

99 *Margin*: Bartholomaeus Scultetus. Cf. Scultetus (1578), Hellman (1944, 206–216), Jerratsch (2020, 243–253).

Thaddaeus Hagecius[a] gewesner *Caesareus Medicus*, sonsten ein fürtreffenlicher *Mathematicus*, wie solches sein *erudita*, und zu lesen wol würdige *Dialexis de Stella anni 1572*. bezeugt, diser im tractat vom Cometen *anni* 1577. vergißt sein selbs, und was er *Dialexi*, p. 64. strafft, im selben wirt er selbs straffbar, dan er nemmt die zeit zwischen seinen zweyen observationen nur 18. minuten einer stund, und bringt hiemit den Cometen ein wenig mehr als 8. *semidiametris terrae* über das *centrum terrae*.

Georgius Busch[b] *cap*. 1. *et* 3. lociert eben disen Cometen in 24. *semidiam. terrae* hinauf, das weren mehr als 20. tausent teutscher meilen.

Aber *Cornelius Gemma*,[c] nemmet zwischen beeden observationen ein geraume zeit, nämlich ettwas mehr als vier stund, darumb kommet er vil näher zu dem zweck, dann er setzt eben disen Cometen *anni* 1577. über den Mon, ins *Mercurii sphaeram*. Nun aber ist er (wie in meinem *scripto* von selbigem Cometen erwisen, und Tycho Brahe[d] kundtschafft gibt) in *sphaera Veneris* gestanden. Irret also gar nichts, das ettliche *Mathematici* die ob erzehlte Cometen nicht für aetherisch, sonder für elementarisch haben ausgerüffet.

Darnach,[e] man findt auch bey den *Historicis* und *Philosophis* noch mehr Exempel von Cometis, deren ettliche in *aethere* sind observiert worden. | Aber hievon soll in eim besondren Tractat, ob Gott will, ausführlich gehandlet, und angezaigt werden, das vil mehr zu schließen were, **Es seye kein Comet jemals in *regione elementari*, sonder alle samptlich in *aetherea regione* erschinen.**

Doch kan Ich alhie nicht umbgehen, fürzubringen, was von dem großen Cometen, deßen Aristoteles[f] in *Meteorologicis* in sonderhait gedencket, zu halten seye.

Aristoteles[g] *lib*. 1. *cap*. 6 *et* 7. *Meteor*. in dem er der alten *Philosophorum* traditionen und fürgeben widerlegt, und hingegen das die Cometen nit im Himmel, sonder im Lufft generiert, angezündt und bewegt werden, beweiset, thut er in sonderheit meldung eines großen Sterns oder Cometen, welcher zu seinen zeiten gesehen ist worden. Wer wolte da nit darfür halten, das, wan jemals ein

---

[a] *Margin*: Thaddaeus Hagecius. [b] *Margin*: Georgius Busch. [c] *Margin*: Cornelius Gemma. [d] *Ms.*: *Bahe*. [e] *Margin*: 3. Es sind auch vor langst, Cometen droben im himel observiert worden. [f] *Deleted*: *lib. 1. cap. 6 et 7*. [g] Margin: 4. Ein Comet zu Aristotelis zeitten ist auch in Aethere gestanden.

100 *Margin*: Thaddaeus Hagecius. The comments on Scultetus and Hagecius are absent in the rough copy (p. 22).

101 Maestlin refers to Hagecius (1574).

102 Cf. Hagecius (1578). On this work and on Maestlin's criticism, see Hellman (1944, 184–204, especially 201f.).

Thaddaeus Hagecius,[100] former Imperial Physician and otherwise an admirable mathematician, as witnessed by his erudite and entirely worth reading *Dialexis de stella anni 1572*,[101] forgets himself in his treatise on the comet of 1577; and what he denounces on page 64 of the *Dialexis* amounts to denouncing his own person, for he takes only 18 minutes as the time between his two observations, and thereby removes the comet by little more than 8 earth radii from the center of the Earth.[102]

Georgius Busch[103] in Chapters 1 and 3 locates the same comet 24 earth radii up, that is, by more than 20,000 German miles.

Cornelius Gemma,[104] however, takes a long time between his two observations, namely slightly more than four hours, and therefore comes much closer to the mark, for he situates the same comet of 1577 above the Moon in the sphere of Mercury. To be sure, it was situated in the sphere of Venus (as shown in my work on the same comet[105] and Tycho Brahe[106] makes known). Yet it is still true that several mathematicians have declared the above comets not as ethereal, but as elementary.

In addition,[107] we may also find among the historians and philosophers further examples of comets, many of which were observed in the ether. | But this 30
shall be dealt with in detail in a special treatise, God willing, where it will be shown that we have much more [reason] to conclude that **no comet has ever appeared in the elementary region, but in each and every instance in the ethereal region.**[108]

Here, I cannot avoid, however, bringing up what should be held about the great comet that Aristotle, in particular, thought of it in the *Meteorologica*.

In Book 1,[109] Chapters 6 and 7 of the *Meteorologica*, in which Aristotle refutes the traditions and treatments of ancient philosophers, while at the same time establishing that comets are generated, kindled, and follow their course not in the heavens but in the air, he takes special note of a great star or comet that was seen in his own times.[110] Who would then not wish to hold that, whenever a

103 *Margin*: Georgius Busch. Cf. Busch (1577), Hellman (1944, 225–233).

104 *Margin*: Cornelius Gemma. Cf. Gemma (1578, 36). This comment on Gemma is absent in the rough copy, p. 22.

105 Maestlin (1578, 38, 40, 54, 56), Hellman (1944, 177–182).

106 Brahe (1588, 159–161), Hellman (1944, 182), Christianson (2020, 91).

107 *Margin*: 3. Long ago, comets were observed up in the sky.

108 This future treatise was never written.

109 Margin: 4. A comet at the time of Aristotle also took place in the ether.

110 Aristotle (1984), *Meteorology*, I, 6, 343b 1–4, 18–25; 7, 344b 34–345a 1. On this comet, dated

Comet were im Lufft erschinen, deßen man glaubwirdige kundtschafft hette, so were es freylich diser Comet gweßt, und were Aristoteles one zweifel durch solche gelegenhait verlaittet worden, dise sein opinion und meinung wider die alte *philosophos* zu bestreitten? Aber wan man selbigen Cometen, wie er von *Aristotele* selbs, und von andern beschriben wirt, ettwas fleißigers beschawet, so finden sich an ime vil wichtigere aigenschafften eins himlischen, als eins Irdischen oder Elementarischen Cometen. Darumb kan und mag[a] aus des *Aristotelis* kundtschafft und aygner Aussag wider ine *Aristotelem* selbs gar wol geschloßen werden, das selbiger Comet nit im Lufft, sonder im hohen himmel erschinen seye.

Diser Comet, sagt Aristoteles, ist in grimmigen kalten winter bey schönem hellem himmel, auf den Abendt erschinen, und zwar hatt man ine den ersten tag nit gesehen, dann er ist vor der Sonnen undergangen [gleich wie man jetzigen unserm Cometen, dieweil er anfänglich früe aller erst nach der Sonnen ist aufgangen auch nit hatt sehen könden] folgenden tag aber hatt er sich ein wenig sehen laßen, dann er war nach der Sonnen undergang umb ettwas, doch
31 nit weit da|hinden gebliben, darumb er bald nach ir auch ist undergangen. Er hatt sich aber bey dem *Aequinoctial* circkel herfür gethon, und seine Strahlen oder Striemen bis ins drittail des himels hinauf, gleich wie ein Wald, gestrecket. Deshalben ward er genennt ὁδός, *via*, ein Weg, oder Straß [one Zweifel, wie wir *Galaxiam*, oder *circulum Lacteum*, nennen die Hörstraß,[b] oder S. Jacobs Straß]. *Diodorus Siculus lib.* 15. nennet in, ein große brinnende fackel, so wegen seiner gestalt ein fewriger Balck, genennet worden, und so hell geschinen, das er auf der Erden ein schatten gemacht, wie der Mon. *Seneca lib.* 6. *Natur. quaest.* haißt in ein[c] große fewrige Saul, und *lib.* 7. es seye am himmel ein großes und ungewohnliches liecht, wie ein großer Balck erschinen, welchs vil tag geleuchtet hab, doch spricht er, Aristoteles habs für keinen wolcken, sonder für ein Cometen erkennet. Diser Comet, sagt Aristoteles weitter, hatt (in nachfolgenden tagen, dann er hatt ettlich vil tag geschinen) sich bis hinauff zu des Orions gürtel gezogen, alda ist er wider verloschen.

---

[a] *Margin: kan und mag.* [b] Heerstrasse, *via regia.* [c] *Added above the line*: ein. "In" = ihn, as above in the reference to Diodorus "nennet in" = "nennet ihn".

---

now in 371 BC, see Pingré (1783, 259–263), Kronk (1999, i, 4–5). For an earlier date (373 BC), see p. 33 below.

111 The Pythagoreans, Hippocrates of Chios and Aischylos, discussed in *Meteorology*, I, 6, 342b 30–343a 35.

112 Aristotle (1984), *Meteorology*, I, 6, 343b 18–23: "the great comet we mentioned before appeared to the west in winter in frosty weather when the sky was clear, in the archonship of Asteius. On the first day it set before the sun and was then not seen. On the next

comet about which we might have credible knowledge should appear in the air, this comet would, in fact, be so, and Aristotle would, no doubt, not have been induced by such an opportunity to dispute the ancient philosophers[111] on the basis of his own conviction and opinion? But if we take a slightly closer look at the same comet as it is described by Aristotle himself and others, we may find in it far more significant properties of a celestial, rather than an earthly or elementary, comet. Consequently, we can and may firmly conclude that, by turning the knowledge of Aristotle and his own testimony on himself, the same comet did not appear in the air but in the upper heaven.

This comet, says Aristotle,[112] appeared one evening by a bright and clear sky during a cold and grim winter, though no one saw it the first day, in fact, for it set before the Sun [just as no one could see our present comet, for it originally rose first thing in the morning right after the Sun], but it became visible for a short time the following day, since it lingered a short while after sunset, though
not long |, and for that reason set soon after it as well. Yet it extended to the 31
equinoctial circle, and its rays or streaks spanned up to a third of the sky like a forest. That is why it was called a path, way, road or route [no doubt the same way that we refer to the galaxy or Milky Way as the Military Road[113] or St. Jacob's Street]. Diodorus Siculus in Book 15 calls it a great blazing torch known according to its form as a fiery beam, which shone so bright that it cast a shadow on the Earth like the Moon.[114] Seneca in Book 6 of the *Naturales quaestiones* calls it a great column of fire,[115] and [he says] in Book 7 that it appeared as a great and extraordinary light in the sky, like a fiery beam, which shone brightly for many days, but adds that Aristotle did not identify it as a cloud but as a comet.[116] Aristotle further says that this comet proceeded (in later days, for it shone for quite a number of days) all the way up to the belt of Orion, where it vanished again.[117]

---

day it was seen, being ever so little behind the sun and immediately setting. But its light extended over a third part of the sky like a band so that people called it a *path*". This and the following paragraph are absent in the rough copy.

113 Heerstrasse, *via regia*. See Grimm, *Deutsches Wörterbuch*, s. v. *Heerstrasze*: "Auch die milchstrasze heiszt *heerstrasze* [...]; für die gegenwart ist der name noch aus Schwaben [...] bezeugt".

114 Diodorus Siculus (1933–1967, XV, 50, 2–3): "There was seen in the heavens during the course of many nights a great blazing torch which was named from its shape a 'flaming beam' [...]. At any rate this torch had such brilliancy, they report, and its light such strength that it cast shadows on the earth similar to those cast by the moon".

115 Seneca (1972), *Naturales quaestiones*, VI, 26, 3: "immense column of fire".

116 Seneca (1972), *Naturales quaestiones*, VII, 5, 4: "Aristotle says that this was not a Beam [*trabem*] but a Comet".

117 Aristotle (1984), *Meteorology*, I, 6, 343b 24–25: "This comet receded as far as Orion's belt and there dissolved".

Aus erzehlten[a] umbständen ist klährlich zu vernemmen, erstlich, das selbiger Comet seinen gang *motum proprium*[b] bey oder neben den *Aequinoctial* circkel, von Nidergang[c] *secundum signorum ordinem*, schlechts gegen Aufgang ghabt hatt, und sich nicht gegen Mitternacht, sonder ein wenig gegen Mittag gelendet. Dann er ist erstlich im zaichen des Schützes oder Steinbocks, da auch die Sonn, als zu winters zeit, war, gestanden, von dannen an dem ort des himmels, da der *Aequinoctialis* undergeht, sich aus der Sonnen schein herfür gethon, und gleich stracks wegs dem *Orioni* zu gezogen, bis er seine Gürtel erraicht. Da er dann im 20. oder 24. grad des Stiers, über den *Aequinoctial* gegen mittag[d] (an welchem ort die Stern selber gürtel dazumal gweßt) wider vergangen. Ist also diser Comet vom Schützen, durch den Steinbock, Waßermann, Fisch und Wider[e] geloffen, und ehe er verschwunden, bis in 20. oder 24. grad des Stiers komen.

32 Darnach ist hieraus leichtlich abzunemmen, das diser Cometen lauff gar richtig, ordenlich und wol proportioniert gweßt sey, seye auch von anfang bis zu end kein ungleiches umbschwirmen daran gespürt worden, sonsten würde es Aristoteles nicht verschwigen haben, seitenmal es ihm in seinen kram, die Cometen, wegen unordenlichen lauffs, aus dem himmel abzuschaffen, und herunder in die Elementarische unordnung zu versetzen, gar taugenlich gweßt wer. Hie frag Ich: „Solte aus disem des Cometen richtigen und ordenlichen gang *Aristoteles* aus seinen aignen worten, nicht wider sich selbs argumentieren, schließen, und also sagen: Diser Comet hatt in allem seinem gang oder lauff nit den Irdischen rauchenden Düfften, und Dämpfen, oder was auch sonsten aus der Erden aufrüechet, ja auch mit aller Elementarischer unbeständigkait, nichts gemein ghabt, sonder er ist in richtiger ordnung und gewißer proportion, aller dings wie die Stern im hohen Himmel, fort geschritten, und mit den selben sich wol verglichen; darumb ist für gewiß zu halten, Er habe nit hie unden bey den Elementen, sonder droben im Himmel in *aliqua Sphaera coelesti* die Zeit seiner Erscheinung zugebracht". Bleibt also auch diser des *Aristotelis* Comet gleich so wol droben im himmel, als die Cometen, welche, darvon droben gehandlet worden, zu unsern lebzeitten erschinen seind.

Wan aber[f] (die zeit zu wißen) diser Comet erschinen seye, zaigt *Aristoteles* am gemeldtem ort an, nämlich zur Zeit des Erdbidems und waßer güsen

---

[a] From here until the end of the chapter the text coincides, apart from several additions, with pp. 22–23 in the rough copy. [b] Margin: *motum proprium*. [c] *Deleted*: gegen Aufgang. [d] *Margin*: über den Aequinoctial gegen mittag. [e] *Deleted*: durch. [f] *Margin*: Die Zeit des obgemeldten Cometen.

According to the above circumstances, it can be clearly understood that, first, the same comet possessed a proper course of motion[118] on or near the equinoctial circle, from the West according to the order of the signs, straight toward the East, and did not turn toward the North but slightly toward the South. For it was first situated in the sign of Sagittarius or Capricorn, where the Sun also was since it was winter, and from there emerged from beneath the light of the Sun to the place in the heavens where the equinoctial line passes below, and immediately proceeded toward Orion until it reached his belt. From there, it then arrived at 20° or 24° Taurus, above the equinoctial line, toward the South[119] (the same place the stars in the belt were at the time). And so the comet passed from Sagittarius through Capricorn, Aquarius, Pisces, and Aries, and first disappeared when it reached 20° or 24° Taurus.

Thus, we may readily deduce from this that the course of the comet was completely regular, uniform, and well proportioned, and that it exhibited from beginning to end not the slightest sense of sudden variation, otherwise Aristotle would not have kept silent about it, since it would have been entirely appropriate for him in his endeavor to abolish comets from the heavens on the basis of their irregular course and transfer them down to elementary disorder. Here, I ask: "Should Aristotle not conclude, from the regular and uniform course of the comet, that his very own words contradict him, and thus say that the comet, over its entire course or path, possessed nothing in common with the earthly fumes, vapors or anything else arising from the earth, nor with any type of terrestrial mutability, but progressed in perfect order and precise proportion, in every way like the stars in the upper heaven, and compared with them favorably;[120] and, therefore, that we may hold for certain that it spent the duration of its appearance above in the heavens in some celestial sphere?" Thus, even the comet of Aristotle lies as far above in the heavens as those comets which we dealt with previously that appeared in our lifetime. 32

When[121] the comet appeared (to ascertain the time), Aristotle indicates in the above passage,[122] namely at the time of the earthquake and sea storm

118 Margin: *motum proprium*.

119 *Margin*: über den Aequinoctial gegen mittag.

120 See Chapter 4, pp. 16–17, on proportion and regularity in cometary motion as an argument for the celestial location.

121 *Margin*: The time of the reported comet.

122 Aristotle (1984), *Meteorology*, I, 6, 343b 1–3, 19–20: "The great comet which appeared at the time of the earthquake in Achaea and the tidal wave [...] in the archonship of Asteius".

33 oder Auslauffs des Meers in *Achaia*[a] |[b] als *Aristaeus* zu Athen Archon, oder oberster Regent war, *Pausanias* in *Achaicis lib.* 7. et 8. und *Diodorus Siculus lib.* 15. nennen in *Asteum*, und seye Archon gweßt *Anno* 4. *Olymp.* 101. (vor *Christi* geburt 373. Jar) zway Jar, ehe die Lacedemonier die Schlacht bey *Leuctra*, und das Regiment, welches sie in Griechenland bey 500. Jar ghabt, verlohren haben (dazumal war *Aristoteles* bey 12. Jar alt, dann er ist, wie *Diog. Laert.* sagt, geborn *Anno* 1. *Olymp.* 99, vor *Christi* geburt 384.). Zur selben zeit waren in *Peloponneso* uberaus große Erdbidem, und jamerliche Waßergüsen, welche ettliche[c] Stätt und Landtschafften verderbten, dergleichen in vorigen Zeitten niemals erhört worden. Dann über dis, das die hauptstatt Sparta,[d] oder Lacedaemon durch grausame Erdbidem vil mal also zerrittelt worden, das schier kein haus auffrecht oder unzerschellet stund, so sind in sonderhait in gemeldtem jar in der Landtschafft *Achaia* am Corinthischen Meer zwo fürneme Stät gar undergangen. Die eine *Buris*,[e] so 40. *stadiis*, das ist, ein große teutsche meil wegs vom Meer, auf einem Berg gelegen. Dise ist in einem Erdbidem also zerfallen, das kein Mensch, on was in Kriegen, oder sonsten anderer Handtierung wegen, nit zu haus gweßt, hatt könden darvon komen. [Ein gleicher jämerlicher *casus* hatt sich Zinstag, den 25. Augstmonats in vergangnem 1618. Jar begeben, da in Grawbündten, ein großer Berg ab im selbs geschliffen, und ein stattlichen Marcktflecken, Plours,[f] ein halbe meil von Cläff, augenblicklich überfallen. Die Gebew, Kirchen und gewaltige schöne Paläst fort geschoben, selbige sampt allen Inwohnern, deren an Mann, weibern und kindern über 2000, ettliche
34 sagen über drey tausent, erbärmlich erschlagen, und zumal also begra|ben, das man kein einig merckzaichen einiges Gebews oder wonung, so da gestanden were, sehen kan].

Die ander Statt war *Helice*, 12. *stadiis*, oder über ein viertel einer teutschen Meil von Meer, die schönste und lustigste Statt in *Achaia*. Dise ist eben in disem Erdbidem im grund so tieff versuncken, das gleich von stund an das ungestümme Meer darüber gewallet, alles was darinnen an Menschen und vich erseuffet. Und dies ist geschehen bey nacht, da in der finstere niemandt dem

---

[a] *Deleted:* einer L dises elenden und erbärmlichen jamers Achaiae eine Landtschafft in Peloponneso, dergleichen in Griechenland niemals erhört worden. [b] Here the fifth sheet begins. [c] *Margin*: ettliche. [d] *Margin*: Sparta durch Erdbidem ubel zerschittelt. [e] *Margin*: Zwuo Stätt in Achaia versuncken und überschwemmet. [f] *Margin*: Plours in Grawbindten undergangen.

---

123 Pausanias (1918–1935), VII, xxiv, 6–12 and xxv, 4–9; VIII, x, 5; Diodorus Siculus (1933–1967, XV, 48, 1; 50, 2).

124 Diogenes Laertius, *Lives of the Philosophers*, V, 9.

125 *Margin*: Sparta shattered by earthquakes.

or tidal wave in Achaea |, when Aristaeus was archon or highest regent at 33
Athens, known by Pausanias in Books 7 and 8 of his account of the Achaeans
and Diodorus Siculus in Book 15 as Asteus, who became archon in the fourth
year of the 101st Olympiad (373 BC), two years before the Spartans lost the bat-
tle of Leuctra, along with their hegemony in Greece they had held for 500
years[123] (that is, when Aristotle was twelve years old, since he was, according
to Diogenes Laertius, born in the first year of the 99th Olympiad, in 384 BC).[124]
At the same time, throughout the Pelopponese there were great earthquakes
and dreadful tidal waves that ruined several cities and regions, the likes of
which had never been heard of in former times. In addition, the capital city
of Sparta[125] or Lacedaemon was battered so many times by brutal earthquakes
that scarcely any houses stood erect or undamaged, and in the same year two
prominent cities in particular in the region of Achaea on the Corinthian Sea
disappeared entirely. One was Bura,[126] which lay on a mountain 40 stadia or
one long German mile away from the sea. This city collapsed in an earthquake
in such a way that no man, unless he had been away from home in battle or
on some other form of business, could have survived.[127] [A similarly dreadful
case occurred on Tuesday, 25 August, last year (1618) when a great mountain,
in Graubünden, suddenly sank and Plurs [Piuro],[128] a handsome market town
half a mile from Cläff [Chiavenna], fell at once below the surface. Buildings,
churches, and a number of beautiful and prominent palaces [were] carried
away, along with their inhabitants, of which more than 2,000 men, women, and
children, many say more than 3,000, [were] miserably slain or buried | in such 34
a way that one cannot see even the slightest sign of a building or dwelling place
that once stood there.][129]

The other city was Helice, 12 stadia or more than ¼ German miles from the sea, the most beautiful and charming city in Achaea. This city sank down so deeply in the same earthquake that the impetuous sea surged over it at once, submerging everything in it, including the people and their livestock. And it occurred at night, so that in the dark no one could come to the aid of another

126 *Margin*: Two towns in Achaia drowned and flooded. The two towns are Bura and Helice, as mentioned in Pausanias. Cf. note 123 above.

127 Diodorus Siculus (1933–1967, XV, 48, 1–3).

128 *Margin*: Plours in Graubünden disappeared. Cf. Habrecht (1618, 60): "So ist uns allen nun mehr bekand wie weyland die Statt Pleurs im Schweitzergeburg mit allen ihren Inwohnern uberfallen, und augenblicklich von einem grossen Sandberg vergraben worden". The two towns are Piuro (Plurs) and Chiavenna (Cläff o Cläven), then belonging to Switzerland and currently to Lombardy (Italy) near the border with Switzerland.

129 This parenthesis is an addition, absent in the rough copy (p. 23).

andern zu hilff kommen, noch selbst entrinnen kondt. *Ovidius lib.* 15. *Metam.* und Seneca *lib.* 6. *et* 7. *Quaest. Natur.* zum sechsten mal,[a] *Plinius lib.* 2. *cap.* 92. schreiben, es seyen beede Stät zu gleich überschwemmet und erseufft worden. Es haben hernach diejenige, so droben hingeschiffet, die Mauren, Thürn und Gebew underm Waßer gesehen, bis hindernach der Schleim sie bedecket hatt.

Plinius in gemeldtem, und ettlichen vorgehenden Capitteln wie auch *Seneca lib.* 6. erzehlen noch mehr solcher Exempel, und erbärmlicher Zuständ.

[Solches[b] ungestümmen Auslauffens, oder überfahls des Meers ist man sonderlich im Niderland fast gewohnt, dann hiedurch werden die ort, Inseln und land mehrmals verendert, das bisweilen eins zu, das ander abgenommen, bisweilen eine undergangen, die ander aus dem waßer sich wider herfür gethon. Aber der geringern flotten einfahln (welche die Einwohner selbs, ohn wer darmit getroffen wirt, nit achten) zu geschweigen, Es schreibt Ludwig Guicciardin, in der Beschreibung des Niderlandts,[c] das von *Anno* 1421. her (on Zweifel bis zu seiner Zeit, als er solchs geschriben bis *Anno* 1566.) die Wahal (ein Arm des Reins, also genandt), und die Moß, sampt dem Meer, bey Dordrecht in Hollandt, ein gantzen *tractum*, so zu vor mit Brabandt ein Landt geweßt, übergoßen,[d] mit 72. Dörffern, und darinn über hundert mal tausent Menschen, sampt | allem irem gutt elendiglich ertrenckt haben. Also im Jar *Christi* 1530. und 1532.[e] haben die grausame Sturmwind[f] das Meer also wüttend gemacht, das es an vilen orten die Dämme mit gewaldt zerrißen, von Suidbeueland, ein Insel in Seeland, nit vil weniger als das halbe thail,[g] das Nortbeueland, ein Insel zu nächst darbey, gantz und gar zerstört, die Stätt, Dörffer, Höfe, Menschen und vichs mit waßer bedeckt und ertrenckt, das man nichts darvon, als ettwan die spitzen der Thürnen hatt sehen könden. Eben zur selben zeit ist auch Duvelandt, oder Dullaert,[h] ein Insel in Seelandt, mit vil leutten und vich von grimmigen Meers wellen überfallen und bedeckt worden. Daher ernandte ort genennet werden die vertrunckene Land. Welche ort samentlich durch erschrockliche Erdbidem also im grund sind versunken, seitenmal das Meer von inen nit wider, seinem gebrauch nach, ist abgeloffen, sonder stehen gebliben].

---

[a]*Margin*: zum sechsten mal. [b]*Margin*: Gleichförmige erschröcklich Zuständ im Niderland. [c]*Margin*: pag. 207. [d]*Deleted*: und. [e]*Margin*: pag. 244, 245. [f]*Ms.*: Sturwind. [g]*Deleted*: mit Stetten, Dörffer. [h]*Margin:* pag. 239.

---

130 Diodorus Siculus (1933–1967, XV, 48, 1–3). The previous parenthesis has broken the continuity of the passage regarding the old Greek cities.

131 Ovid (1984), *Metamorphoses*, XV, 293–295.

132 *Margin*: For the sixth time. Cf. Seneca (1972), *Naturales quaestiones*, VI, 23, 4; VI, 25, 4; VI, 26, 3; VI, 32, 8; VII, 5, 3–4; VII, 16, 2.

nor escape themselves.[130] Ovid in Book 15 of the *Metamorphoses*,[131] Seneca six times in Books 6 and 7 of the *Naturales quaestiones*,[132] and Pliny in Book 2, Chap. 92[133] write that both cities were submerged and swallowed up at the same time. Those who later sailed through the area saw the buildings, towers, and walls below the water until they began to be covered by mud.

Pliny in the above chapter and several previous ones,[134] as well as Seneca in Book 6,[135] tell of many such other examples and their dreadful conditions.

[Particularly in the Netherlands,[136] one is practically accustomed to such a vehement eruption or overflow of the sea, since islands, lands, and localities throughout the region often shift, some expanding while others contract, some sinking below the surface of the water while others emerge again from it. Without mentioning the sinking of lighter sea vessels (which local residents do not pay any regard to, unless they are forced to face it themselves), Lodovico Guicciardini writes in his account of the Netherlands[137] that from the year 1421 (up to his own time, no doubt, when he wrote his account in 1566), the Rivers Waal (a branch of the River Rhine, thus named) and Maas, together with the sea, flooded near Dordrecht in Holland across a vast expanse of land made up of 72 villages that had once belonged to Brabant, mercilessly drowning more
than 100,000 men along with | everything they owned. In the years 1530 and 35
1532,[138] a fierce wind storm stirred up the sea so strongly that it tore down the dams in many places, destroying no less than half of South Beveland, an island in Zeeland, and completely demolishing an island next to it, North Beveland; towns, villages, farms, men, and livestock were covered and drowned by so much water that no one could have seen any sign of them, except for perhaps the tops of towers. At the very same time, Duiveland or Dullaert,[139] another island in Zeeland, was also overwhelmed and submerged by brutal sea waves. As a result, the above sites are now known as sunken lands. All of these places have sunk to the bottom through terrible earthquakes because the sea has not retreated from them again, according to its custom, but rather remained in place].

---

133 Pliny (1938–1952), II, xciv, 206; for the numbering of the chapter, see note 8 above.

134 Pliny (1938–1952), II, lxxxvi, 200—xciii, 205.

135 Seneca (1972), *Naturales quaestiones*, VI, 26, 3 and 32, 8.

136 *Margin*: Similar terrifying conditions in the Netherlands. This parenthesis is a new addition which is absent from the rough copy (p. 23).

137 *Margin*: pag. 207. Cf. Lodovico Guicciardini (1580, ccvii).

138 *Margin*: pag. 244, 245. Cf. Lodovico Guicciardini (1580, ccxliv, ccxlv), who dates it to 1532.

139 *Margin*: pag. 239. Cf. Lodovico Guicciardini (1580, ccxxxix), where the date of 1530 is given.

Hie ist aber bey allen solchen schwähren und kläglichen Zuständen auch dis wol zu mercken, das vor denselben[a] alle mal eintweder erschrockenliche Cometen, oder andere merckliche und gedenkwürdige wunderzaichen, als von Gott gesandte Vorbotten, vorher sind gangen.

[a] *Margin:* auch dies … vor denselben.

It is also worth noting here that, before each and every one of these hard and difficult circumstances, either frightful comets or other memorable and remarkable marvels [*Wunderzeichen*] have always preceded them as harbingers sent by God.

CAPUT VIII

Wa es dem *Aristoteli* und andern *Philosophis*, von den Cometen anderst zu halten, als es sich befindt, gemangelt habe.[a]

Dieweil[b] disem also, nämlich, das nit allein diser unser *Anni* 1618. Comet, wie *cap.* 6. aus der *parallaxi* erwisen, sonder auch noch andere mehr, als in sonderhait des *Aristotelis* Comet, wie in vorgehendem *cap.* 7. angezaigt worden,
36 nit im | Lufft, sonder im Himel gweßt; möchte man nit unbillich fragen, wie kompts das *Aristoteles*, welcher doch Sinnreich gnug gweßt, und auch andere fürtreffliche und scharpfsinnige *Philosophi* solches nit eben so wol als jetzt zu unsern Zeitten,[c] gemerckt, sonder so grob übersehen haben? ja das deren vil es noch heutigs nit erkennen wöllen?

Antwort. 1. Under den alten *Philosophis* waren nit wenig, welche von den Cometen nit, wie Aristoteles, hielten, sonder lehrten, das die Himmlische oder Aetherische Natur mit nichten aller Verenderung befreyet seye, und, das die Cometen daselbsten entspringen, habens aber aus rechtem grund nit gewußt zu erweisen.[d] Ein zimliche anzahl derselben philosophen erzehlen Aristoteles *lib.* 1. *cap.* 6. *Meteor. Plutarchus lib.* 3. *cap.* 2. *de placit. philos. Seneca lib.* 7 *quaest. Natural.* Als, da waren *Anaxagoras, Democritus, Pythagoras, et quidam in Italia dicti Pythagorici, item Hippocrates Chius, eiusque discipulus Aeschilus, Diogenes, Artemidorus, Apollonius* (*qui Chaldaeos idem statuere dicit, Epigenes tamen contrarium testatur*) *Zenon, ipse etiam Seneca*, etc.

2. *Aristotelem* betreffend, so ist er wol zu entschuldigen. Dann wie er darhinder kommen seye, ist droben *cap.* 4 et 5 angezaigt worden, nämlich, da er die widerige meinungen zu entschaiden, solte, seiner selbs aignen lehr nach, zu vörderst nach dem ὅτι gesehen, und aus rechtem aigentlichem grundt erkundiget haben, wie hoch einer oder der andere Comet ob der Erden gestanden were, und als dann hette er nach dem διότι, waher und waraus sie generiert werden, nachsinnen mögen:[e] dieweil es aber eben an disem ort zuzukomen ime gefählet hatt und auch die obbenandte *Philosophi* ire gutte sach selbs nit kundten *demonstrativè*, von dem rechten ὅτι anzufangen, beweisen:[f] also hatt er diser

---

[a] *Deleted:* Und das nahe und nahe vil künsten und sachen erkundigt seind worden, darvon man vor alten zeiten nichts gewüßt hatt. [b] The first pages of this chapter, until the mention of Aristarchus on page 38, have undergone a great many changes with respect to p. 24 in the first writing. [c] *Margin*: als jetzt zu unsern Zeitten. [d] *Margin*: habens … zu erweisen. [e] *Margin*: mögen. [f] *Added at the bottom*: und auch die … beweisen.

---

140 The first pages of this chapter, until the mention of Aristarchus on page 38, have undergone a great many changes with respect to p. 24 in the first writing.

Chapter 8
What Aristotle and Other Philosophers Might Have Been Missing That Led Them to Think about Comets the Wrong Way

Given the above,[140] namely that not only has our present comet of 1618, as proven by parallax in Chap. 6, but also many others, especially the comet of Aristotle, as shown in the previous chapter, not been in the | air, but in the 36
heavens; might one not unreasonably inquire how is it that Aristotle, who was sufficiently well endowed with reason, along with other astute and eminent philosophers, did not realize this as well as [we do] in our own time, but overlook it so crudely? And that many of them still do not wish to admit it today?

Answer. 1. Among the ancient philosophers, there were many who did not think the same thing as Aristotle about comets, but taught that celestial or ethereal nature was by no means free from any form of change and that comets found their origin there, they just did not know how to prove it on the proper grounds. Aristotle in Book 1, Chap. 6 of the *Meteorologica*, Plutarch in Book 3, Chap. 2 of *De placitis philosophorum*, and Seneca in Book 7 of the *Naturales quaestiones* recount a fair number of these philosophers. They include Anaxagoras, Democritus, Pythagoras and those in Italy who were called Pythagoreans, as well as Hippocrates of Chios and his disciple Aeschylus,[141] Diogenes,[142] Artemidorus, Apollonius (who claims that the Chaldeans held the same view, while Epigenes testifies to the contrary), Zeno,[143] Seneca himself, and others.

2. As for Aristotle, he may certainly be forgiven. For how he arrived at his opinion is shown above in Chapters 4 and 5, that is, since he had to decide between opposing views he should, according to his very own teaching, first have looked after the *quid*, and inquired on truly proper grounds how high one comet or another was situated above the Earth, and only then could he have contemplated the *propter quid*, what from and whence they are generated:[144] however, since he failed to arrive even at this place, and the above philosophers[145] could not prove their case themselves by demonstration, beginning from the *quid* in question, he approved not unreasonably of this opinion, which

141 Aristotle (1984). *Meteorology*, 342b 25–36.

142 Plutarch (1516, f. Dii^{v}): "Diogenes crinitas stellas revera esse". Diogenes of Apolonia (fl. ca. 425 BC) is meant.

143 Seneca (1972, VII, 4, 1; 13, 1; 19,1).

144 On this distinction, see p. 18 above.

145 The supporters of the celestial location of comets.

meinung, welche ein beßern schein ghabt, nit unbillich, beyfahl gethon. Das er aber dise sein mainung nit so gar für gewiß und stählin halte,[a] sonder deren selbs nit gnugsam trawe, zaigt er an, da er sich *lib.* 1. *cap.* 7 *Meteor.* deßen beklagt, und austrucklich gnug protestiert. „Wir haltens darfür", spricht er, „was man von den jenigen dingen, die vor unsern Sinnen verborgen sind,[b] reden will, so seye es weißlich oder verständtlich gnug erklärt, wann mans so weit, als es
37 müglich est, bringet. Nun mag man | wol darfür halten, das es auch alhie, was (von den Cometen) zu sagen ist, also beschaffen sey", etc.

Mit diser protestation, welche *Aristoteles*, da er von den Cometen handlen will, an statt einer Vorred vor her setzt, gibt er klärlich zu verstehen, 1. von den Cometen zu discurrieren, solte der Anfang *à sensibilibus*, das ist, was das Gesicht (dann kein anderer *sensus* hatt hie platz) begreiffen kan, gemacht werden. Er versteht aber *sensibilia propria*, was aigentlich hieher gehört (*sensibilia communia* richten da nicht vil aus). Das seind solche *observationes*, durch welche, wie hoch und weit ein Comet ob und von der Erden stehet, gewis kan ausgerechnet werden, welches wir *doctrinam Parallaxeon* haißen. Wann man dann die höhe erkundigt hatt, und waißt, ob der Comet nur im lufft schwebe, oder sich droben in himel gestellt habe: als dann mag man[c] seiner materi *ex qua* nachgedencken, ob er *ex exhalationibus ex terra*, von den tüfften oder Dämpfen aus der Erden, oder anderstwo her seinen Uhrsprung habe, etc. 2. Bekennet Er, das es eben an solchen *sensibilibus* und observationen ihme[d] gemanglet hab, dann, wie bald folgen wirt, *Doctrina Parallaxeon* ist dazumal noch nit erfunden gweßt. 3. Dieweil nun disem also, und *Aristoteles* den weg, *à priori, sive demonstrativè à causa ad effectum* zu argumentieren, nit wandlen

[a]*Deleted*: ist daraus abzunemmen, das er lib. 1 cap. 7 Meteor. *Margin*: NB. Aristoteles hatt selbs, ob er recht daran sey, gezweifelt, und sich dessen protestiert. [b]*Margin*: (quae sunt ἀφανῆ τῇ αἰσθήσει). [c]*Added above the line*: man. [d]*Added above the line*: ihme.

146 *Margin*: NB. Aristotle himself had doubted whether he was right about it, and bore witness to it.

147 *Margin*: (quae sunt ἀφανῆ τῇ αἰσθήσει). Cf. Aristotle (1984, 344 a 5).

148 Cf. Aristotle (1984, 344a 5–8): "We consider a satisfactory explanation of phenomena inaccessible to observation to have been given when our account of them is free from impossibilities. The phenomenon available suggest the following account of the matters in question".

149 This 'protestation' precedes, in fact, the beginning of Aristotle's exposition of his own doctrine in *Meteorology*, I, 7.

150 On this distinction, see Aristotle (1984c), *On the Soul*, II, 6, 418a 10–13: "Of the first two kinds [of 'object of sense', *sensibilia*] one consists of what is special to a single sense, the other of what is common to any and all of the senses. I call by the name of special object of this or that sense that which cannot be perceived by any other sense than that one and in

had a better appearance. To be sure, he admits that he did not hold his opinion completely for certain or firm,[146] and was not fully confident in it, for he complains about it in Book 1, Chap. 7 of the *Meteorologica* and expressly protests against it as well. "We therefore consider," he says, "what one wishes to say about those things that lie hidden from our senses[147] as explained wisely or in a sufficiently reasonable way, should one progress in this endeavor as far as possible. Now one may | therefore consider that what shall be said here (of comets) has 37
been secured in such a way,"[148] and so forth.

With this protest, which Aristotle presents in place of a preface[149] just as he is about to deal with comets, he clearly wishes to convey that 1. to discourse about comets should be done from the beginning by the senses, that is, what can be grasped by sight (since there is no room for the other senses in this instance). He understands, however, the proper senses [*sensibilia propria*] as that which applies here (the common senses [*sensibilia communia*] do not play much of a part).[150] These are such observations through which we may calculate with certainty the height and distance of a comet above and beyond the Earth, which we call the doctrine of parallax.[151] Once we have thus found the height and know whether the comet was merely floating in the air or situated high above in the heavens, only then may we ponder the matter from which the comet came, whether it originated from terrestrial exhalations, from the fumes or vapors of the earth, or some other source.[152] 2. He indicates that he was lacking even those sensibles and observations, for, as shall soon follow, the doctrine of parallax had not yet been discovered at that time. 3. Given these conditions, Aristotle could not tread down the path of arguing *a priori*, that is,

respect of which no error is possible". See also Melanchthon (1595), *Liber de anima*, f. N3^{v}: "Quid est sensus visus? Est sensus percipiens oculis lucem et colorem, tanquam propria obiecta, et magnitudines, figuras, numerum, motum et situm corporum, tanquam communia obiecta".

151 In accordance with Aristotle and Melanchthon, Maestlin presupposes that the perception of the light of the comet, as the proper object of sight, contains no error, whereas the location (*situs*) as a common object of several senses may be perceived erroneously. However, the application of the doctrine of parallax to the light perceived can establish its location with certainty (*gewis*). For the qualification of parallax as the 'rule of truth' (*regula veritatis*), see Maestlin's statement on the nova of 1604 in Granada (2014, 120): "the doctrine of parallax (which alone in the measurement of distances, without contradiction or exception, can neither deceive nor be deceived, and so without contradiction must be recognized as the rule of truth)".

152 As a consequence, Aristotle inverted the correct methodological procedure, since he departed from the presumed nature of comets and the presupposed immutability of the heavens. Cf. p. 39 below.

könden, dann der weg vorderwerts zu zukommen war dazumal noch verschloßen, so mußt er ja den andern weg, *à posteriori, coniecturaliter ab effectu ad causam*, von hinderwerts durch muttmaßungen gehen und[a] suchen. Doch dieweil mit *coniecturis* und muttmaßungen nichts gewiss kan demonstriert und erwisen werden (*arduum est, de his quae coniecturâ sequenda sunt, aliquid certi promittere.* Es understeht sich, sagt Seneca *lib.* 6. *quaest. Nat.*[b] einr eins hohen wercks, welchs müßlich ist hinaus zubringen, wan er will von einr sach ettwas gewis zu erweisen versprechen, welche er allein durch muttmaßen erraichen muß) so vermeint er doch, es sey der sach gnug geschehen, wan nur nichts unmüglich oder ungereumbtes daraus erfolge, unangesehen es in der warhait
38 schon solche anderst beschaffen sein. 4. Mit welchem | allem *Aristoteles* gnug an tag und zu verstehn gibt, das er dise meinung nit lenger begere zu bestreitten, als bis er eins beßern berichtet werde.

Das aber[c] *Doctrina Parallaxeon*, welche dises alles hette könden richtig machen, zu zeit des *Aristotelis* noch unbekandt gweßt seye, ist daraus zu beweisen. 1. *Aristoteles* in allen seinen Büchern thut deren mit keinem eintzigen wort meldung, thut auch kein Andeuttung darauff. 2. *Aristarchus*, welcher nach *Aristotele* gelebt[d] umb das Jar vor *Christi* geburt 280 (nach ausweisung seiner observation, davon *Ptolemaeus lib.* 3. *cap.* 2. *Almag. Suidas* setzt in über die 100 Jar später, nämlich under *Ptolemaeum Philometorem* König in Aegypten, in *Olymp.* 156. vor *Christi* geburt 154. Jar) waißt *in libello de magnitudinibus et distantiis Solis et Lunae*, auch nichts darvon, sonder braucht frembde, und zu seinem vorhaben übel taugende *argumenta*. Da er doch sonderlich solte die *parallaxes*, wan er ja ettwas darvon gewißt hette, für die hand genomen haben. 3. Ob *Hipparchus*, welcher bey anderhalb hundert jar vor *Christo* gelebt, hiervon gewißt hab, steht bey mir in Zweifel, dann, wiewol *Plinius* und *Ptolemaeus* ine, wie

[a] *Margin*: gehen und. [b] *Margin*: sagt ... Nat. [c] *Margin*: Parallaxeon Doctrina ist zu Aristotelis zeiten noch nit am tag geweßt. [d] *Deleted*: und floriert.

153 Cf. Seneca (1972, VI, 20, 5): "it is difficult to promise anything certain about theories which are based on conjecture".

154 Cf. Aristotle (1984, 344a 5–8), cit. pp. 36–37 above.

155 *Margin*: The doctrine of parallax was still unknown in Aristotle's time. Maestlin affirmed this point in 1578 in order to weaken Aristotle's authority in cometary theory. See Maestlin (1578, 17): "etiam tum parallaxeos indagandae ratio ignota fuit, reservanda fortè ad haec nostra tempora [...]. Dubium non est, si Aristoteles de aethereis per parallaxes inventis certus fuisset (qui enim ante ipsum Cometas in coelum reponebant, quae dicebant, probare non poterant) profectò sententiam conceptam mutasset".

156 The approximate dates of Aristarchus are 310–230 BC.

by demonstration from cause to effect, for the way to proceed forward was still closed at that time, so he had to go the other way, *a posteriori*, conjecturally from effect to cause, moving backwards and seeking a way through speculation. Yet since nothing certain can be demonstrated or proven through conjecture and speculation ("It is difficult to promise anything certain about theories which are based on conjecture." The author of any great work that must be brought to light, Seneca says in Book 6 of the *Naturales quaestiones*,[153] bears these words in mind whenever he wishes to promise to prove something certain about a matter that he must reach purely by speculation), he then supposes the matter to be sufficiently resolved simply when nothing impossible or absurd should come from it, without considering that in truth such things are determined differently.[154] 4. In view of all of the above |, Aristotle makes it sufficiently clear 38
that he no longer wishes to dispute this opinion until he is informed of a better one.

That Aristotle[155] was not familiar in his own day with the doctrine of parallax, which could have cleared up the entire matter, shall now be shown. 1. In none of his books does Aristotle make reference to it with a single word nor express a single hint about it. 2. Aristarchus, who lived after Aristotle around the year 280 BC[156] (as indicated by his [Aristarchus'] observation in Book 3, Chap. 2 of Ptolemy's *Almagest*;[157] Suidas locates him more than 100 years later, namely during the reign of Ptolemy Philometor, King of Egypt, in the 156th Olympiad or 154 BC), is equally unaware of it in his little book, *De magnitudinibus et distantiis solis et lunae*, but makes use of strange arguments poorly suited to his purpose.[158] Precisely here, however, had he known anything about it, Aristarchus should have taken parallax into account.[159] 3. Whether Hipparchus, who lived around 150 BC, was familiar with it remains doubtful to me, for as much as Pliny

157 More precisely Chapter 1. See Ptolemy (1984, 137–139) for the observation of the summer solstice in 279 BC. Maestlin seems to mistake him for the grammarian and critic of the Homeric poems Aristarchus of Samothrace (ca. 220–ca. 143 BC). The encyclopedia Suidas or Suda has an entry for the latter, referring also to his life in the reign of Ptolemy Philometor; it does not mention the astronomer. This reference to Suidas is absent from the rough copy, but already appears on p. 2^{r} of the five slips of paper that represent an intermediary state between the rough copy and the clean copy.

158 On Aristarchus of Samos and his book *On the Sizes and Distances of the Sun and the Moon*, see Heath (1981).

159 The final copy omits the following mention of Archimedes in the rough copy, p. 25: "Also Archimedes vor Christi geburt 210. Jar, wüßte auch ~~von keiner parallaxi~~, sonderlich in *libello de Arenae numero*, nichts von einiger parallaxi".

billich, hoch rhümen wegen seiner sondern geschicklichait, so sucht er doch *in motu et anno Solari et Lunari* ungewißes durch noch ungewißers, da doch *doctrina parallaxeon* ime wol hette helffen könden. Ob auch *Plinius*, schier hundert Jar nach *Christi* geburt, hiervon wißenschafft gehabt, ist zu zweifflen, seitten mal er *lib.* 2. da er vilfältig vons Mons lauff, und der Sonnen fünsternus handlet, gedenkt er der *parallaxeon* nicht. Ist des wegen *Aristoteles* für wol entschuldigt zu halten. Et hatts erkhennt, und bekhennets, waran es im gefählt hab, und, wie angezaigt, protestiert darunder, hatt aber *propter iniuriam seculi*, zur selben zeit im selbs nit zu helfen gewüßt.

3. Gleichergestalt seind auch die andere *Philosophi*, welchen dise *Parallaxeon doctrina* noch zur zeit ist unbekandt gweßt, zu entschuldigen.

39 4. Was soll oder will man aber von denen *Philosophis* sagen, welche zu unsern zeitten disem alten wohn noch so hartnäckig anhangen? Zu verwundern, oder zu beklagen ists, das deren ettlich sich finden, bey denen des *Aristotelis opinio* und meinung, wider und über sein willen und protestieren, so tieff eingewurtzelt hatt, da sie doch nit nur aus den alten *historiis* hören, sonder jetziger zeit handtgreiffig *per parallaxes* überwisen werden, das nun mehr innerhalb 90. Jaren alle Cometen droben im himmel, und nit im lufft, erschinen seind, nichts desto weniger müßen sie inen nur im Lufft sein, und deren ettliche derffen noch,[a] dise ir gefaßte opinion rhümen als[b] ein wolgegründte Lehr, welche alle vernünfftige *Mathematici* und *Physici* also halten. So man nun fragt, Lieber, auf welchen so festen grund ist *doctrina*, das die Cometen aus dem rauch dampf, und faißten, schwer belchen, kleberigen Dünsten, im obern Lufft, generiert werden, so steif und wol gegründet? So wirt geantwortet (doch aus einem *praesupposito*, welches man müßte für gewis annemmen, wanns schon nicht bewisen ist):[c] Dieweil die Cometen nit im himmel[d] sein könden, dann daselbsten gebe es keine Enderungen, darumb[e] so müßens hie unden under des Mons *sphaera* sich bey den Elementen halten, und zwar im obern Lufft, dann deren keiner kompt zu uns hernider. Jetzt, sprechen sie,[f] da man das ort und Losament der Cometen hatt, so seye[g] leucht zu coniecturieren und zu schetzen, waraus sie iren uhrsprung haben, seittenmal in der Elementa-

[a] *Deleted*: nennen. *Added above the line*: deren ettliche derffen noch. [b] *Added above the line*: rhümen als. [c] *Added in the margin*: wanns schon nicht bewisen ist. [d] *Written above the line*: himmel. [e] *Margin*: darumb. [f] *Margin*: sprechen sie. [g] *Margin*: seye. *Deleted*: ist.

160 Pliny (1938–1952), II, ix, 53; x, 57; xxiv, 95. Ptolemy (1984, 131): "Hipparchus, who was industrious and a lover of truth".

161 See Pliny (1938–1952), II, vi, 43- xi, 59.

and Ptolemy rightly pay him high praise for his rare skill,[160] he goes from one uncertainty to a still greater one in his study of the motion and circuit of the Sun and Moon, precisely when the doctrine of parallax could have come to his aid. It is also doubtful whether Pliny, almost 100 years after Christ, possessed any knowledge of it, since in Book 2, where he deals variously with the motion of the Moon and solar eclipses, he does not summon parallax.[161] It is for this reason that Aristotle should be considered completely forgiven. He realized and recognized what he was missing, and, as shown above, expressed his concern over it,[162] but he did not know how to help his own cause on account of the ignorance of the times [*propter iniuriam seculi*].

3. In the same way, we should also forgive the other philosophers to whom the doctrine of parallax was still not known in their day.

4. Yet what shall or should we say about those philosophers who still cling 39
so stubbornly to the same old opinion in our own times? It is astonishing, or regrettable, that there are many of them in whom the opinion and position of Aristotle remains so deeply rooted, against and beyond his own wishes and objections, that even though they learn not only from older accounts but in our own time, more palpably by parallax, that for the last 90 years all comets have appeared above in the heavens and not in the air; nevertheless, they must exist only in the air for these [philosophers], and many of them may still parade this fixed opinion of theirs as a well founded doctrine that all reasonable mathematicians and natural philosophers accept in the same way.[163] So now one asks, dear [philosopher], on what foundation so firm is the doctrine that comets are generated from moist exhalations and heavily laden viscous vapors coalescing in the upper air so soundly and well established? The answer shall thus be (though from a presupposition that one ought to accept for certain even if it has not yet been proven): since comets could not exist in the heavens, for there would be no change there, they must abide down here, below the sphere of the moon among the elements, particularly in the upper air since none of them ever approaches us on earth. Now, they say, since we know the dwelling place of comets, it may be easy to conjecture and estimate from where they have their origin, as every form of change is found in elementary nature: and it

162 See p. 37 above with note 148.

163 After mentioning Hipparchus, Pliny, and Ptolemy as well as the unsound but tenacious adherence of contemporary Aristotelians to the old Aristotelian doctrine, the rough copy introduces the point that the rejection of the sublunarity of comets entails the ruin of physical science. However, in the clean copy this is introduced only on p. 43, in connection with Keckermann.

rischen Natur sich allerlay verenderungen finden: und aber offenbar ist, das aus der Erden und waßern täglich vil Dämpf aufsteigen, deren ettliche zu waßer und regen werden, von ettlichen komen fewr, blitz, donner, stral etc., so seye gutt zu erachten das[a] deren Dämpf ettliche noch höher steigen, und sich samlen im hohen lufft, wie ein berg, bis sie ausgedörret und praepariert werden, als dann werden sie vom Element des fewrs, welches zu nächsten droben ist, und von dem schnellen | umlauff des himmels angezündet, und dieweil der himmel auch des fewrs *sphaeram*, sampt dem obern thail des Luffts, gleichsam mit gwalt mit sich herumb führet, also müßen auch mit selbigem die Cometen herumb fahren, etc.

Dises sind *iucundae Speculationes et Phantasiae*, feine, lustige und holdselige gedancken. Aber, lieber, seind das alles nit nur lauter *praeiudicia* und vorgefaßte Urtheil,[b] im grund aber nur *coniecturae* und muttmaßungen, welche gründtlich nichts beweisen? Dann es haißt ja noch nicht weitter, als, Es ist gläublich, diesem seye also. Aber, sage mir, wer ist hinauf gestigen, und den Cometen dort gefunden, sein[c] stell und ort[d] gesehen, sein Distantz von der Erden abgemeßen? Wer hatt daselbsten gesehen, wie die Dämpf hienach, und so weit hinauf fahren, ausdorren, und angezündet werden? Ist auch jemals ettwar mit dem Cometen umb die Erden herumb gefahren, darbey er gemerckt hatte das der obere Lufft vom himel herumb getriben werde? Aber dis alles, und noch vil anders so disen anhängig, wirt mit dem aintzigen, **Nein**, umbgestoßen und widerlegt, so man sagt, doch auch mit besserm grund erweiset:[e] die Cometen schweben nit im Lufft, sonder im hohen Himmel.

Ja, sprichstu, Wer ist dann bis[f] in himmel, welcher noch höher ist, hinauf gestigen oder geflogen, und die Cometen daselbsten gefunden? Antwort: Droben seind die Laitter, oder flügel gezaigt worden,[g] daran und mit welchen die *Astronomi* hinauff steigen oder fliegen, nämlich *Geometria* und *Arithmetica*. Dise nennet Plato *Alas Astronomiae*, der Astronomi flügel, damit sie bis in himmel, wie hoch auch derselb ist, flieget. Gleich wie daselbsten in der figur, pag. 19, und beygesetztem Exempel, die *Geometria* durch die observierte *angulos*, CAF, 67. grad, und CBF 70. grad, bis zu des Thurns DC spitzen C gestigen oder

---

[a] *Deleted*: seyen. [b] *Ms.* Urthel. [c] *Margin*: dort ... sein. [d] *Added above the line*: ort. [e] *Margin*: doch ... erweiset. [f] *Added above the line*: bis. [g] *Margin*: *Cap. 5*.

---

164 This is a fair and complete summary of the peripatetic theory of comets as resulting from dry exhalations rising to the upper region of the air. See Martin (2011, 5 ff.).

165 The following criticism of the peripatetic doctrine including the reference to Copernicus' demonstration of lunar parallax and the Moon's distance from the Earth (pp. 40–41) appears in draft on fascicle 8b. Cf. p.1 (the one after p. 11) of the five slips of paper that represent an intermediary state between the rough copy and the clean copy.

is evident that many exhalations arise every day from the earth and water, with some turning into water and rain, while others yield fire, lightning, thunder, rays, and so forth, such that it is worth considering that many of the exhalations ascend even higher and assemble together in the upper air, like a mountain, until they are parched and prepared, as it were, to be ignited by the element of fire, which lies just above, and by the swift | circuit of the heavens; and since the 40
heavens move around with them by force, as it were, the sphere of fire, together with the upper part of the air, comets must also move around with them, and so forth.[164]

These are fun speculations and fantasies, fine, pleasant, and pleasing thoughts. But, dear [philosopher], are they not all just a bunch of prejudices and preconceived notions,[165] in essence only conjectures and speculations, which essentially prove nothing? For this amounts to nothing more, in fact, than [saying], "It is believable, so let it be so." But, tell me, who has ascended on high and found the comet there, observing its place and position and measuring its distance from the Earth? Who saw in that same place how the exhalations rose up so far, dried out, and were set on fire? And did he also ever travel anywhere around the Earth with the comet, whereby he would have noted that the upper air is driven around by the heavens?[166] Yet all of this, and many other things that depend upon it, shall be overturned and refuted by a single **No**, so that one may say, but also prove with better reason, "Comets do not float in the air, but in the upper heaven."

"Yes," you say, "but who has climbed or flown all the way up to the heavens, which are even higher, and encountered the comets there?" Answer: we were shown above the ladder or wings[167] on or with which astronomers climb up or fly, namely geometry and arithmetic. Plato calls these the *alas astronomiae*, the wings of astronomy, with which it may fly up to the heavens, no matter how high they are. Just as there in the figure on page 19 and the accompanying example, geometry climbed or flew through the observed angles CAF 67° and CBF 70° to the top C of the tower DC and found that, where lines AC and BC

166 Cf. Copernicus (1543, 6^r). The rejection of the (daily) motion of the upper air is in accordance with the admission of the daily motion of the Earth, which does not extend to this upper region. Maestlin's adherence to Copernicanism, public ever since his collaboration on Kepler's *Mysterium cosmographicum* (1596) is implicit here. See p. 43 below.

167 *Margin*: *Cap. 5*. Cf. p. 22 above. The image of the wings of the soul derives from Plato (cf. *Phaedrus* 246b ff.) and Melanchthon identified them with arithmetic and geometry. Cf. Melanchthon, "Preface to Arithmetic" (1536), Melanchthon (1999, 93): "Consequently, the wings of the human mind are arithmetic and geometry. If someone endowed with an intellect that is not mean attached these to himself, he would easily enter heaven and would wander freely in the heavenly company, and enjoy that light and wisdom".

geflogen, und gefunden, das, da die Linien AC und BC im C zusamen lauffen, faßen ein *angulum parallaxeos* ACB 3. grad groß, deßen *anguli sinus proportio* sich gegen AB helt, wie die *proportio sinus anguli* ABC gegen AC, und hiemit die Lenge der lini AC gemeßen, nit anderst, als hette sie dieselbige Lini vom A, bis ins C, im Lufft mit einer Meßrutten, oder Zollstab abgemeßen: Eben also hat
41 nach Ausweisung folgender figur, pag. 20, *Copernicus*,[a] |[b] die ferne des Mons von der Erden gemeßen. Er fandt aber *angulum* BAC, oder DAC 80. gr. 55 scr. und hat observiert *angulum* DBC 81. gr. 55 scr. Hiemit ist er bis zum Mon selbs hinauf geflogen, und den *angulum* oder *parallaxin*[c] ACB, da beede Linien AC und BC in dem Mon (in der stunde seiner observation) zusamen geloffen, gemeßen und funden, das er war 1. gr. oder 60 scr. Hieraus gab in *Geometria* und *Arithmetica* ferner, gleich wie *anguli parallaxeos* ACB 1. gr. *sinus* 1747. sich helt gegen AB einem[d] *semidiametrum Terrae*; also ist auch *anguli* ABC *sinus* 99006. gegen AC. Auf dise Weis hat *Copernicus* die Lini AC, das ist, wie weit der Mon (daßelbig mal) vom *Centro Terrae* gestanden, gemeßen und funden, das die distants war 56 *semid. terrae* und zway dritteil.

Auf gleichformige weis, wie auch droben cap. 5 gemeldet, gibt *Geometria* onfählbare *demonstrationes*, der Cometen distantz zu rechnen, an die hand,[e] lehret, wie die *parallaxes* zu finden, und zu underschaiden seyen, Als, were ein Comet näher bey der Erden als der Mon, so müßte die *parallaxis* (in vorigem *Schemate*, ACB) umb so vil größer sein. Ist er aber weitter hindan, so muß die *parallaxis* (AMB) auch desto kleiner sein. Hieraus folgt unwidersprechlich (**dann *Geometria* und *Arithmetica*[f] laßen inen nimmer mehr unrecht thun**) dieweil aller Cometen, welche bisher rechtmäßig seind observiert worden, *parallaxes* vil kleiner waren,[g] als sie[h] der Mon gibt, ja an den mehrern theil ist gar kein *parallaxis* gespürt worden (wie alberait von disem letzten *Anni* 1618. Cometen, droben cap. 6 erwisen) so muß gewis folgen, das sie[i] von der Erden[j] nit nur höher stehen als[k] der Mon, sonder im himmel[l] so hoch erhoben sein,[m] das die halbe Erden Dicke, gegen irer höhen,[n] kein *sensibilem proportionem* mehr haben kan, seind als nil, als nichts zu rechnen seye.[o]

Aber alhie wehret sich mit allen kräfften, neben ettlich andern, in sonderhait Bartholomaeus Keckermann, in *Systemate Physico*, und in *Disputationibus*

[a] *Margin: lib, 4, cap. 16 et 17*. [b] Here the sixth sheet begins. [c] *Margin*: oder *parallaxin*. [d] *Margin*: einem. [e] *Deleted*: und. [f] *Deleted*: so fern nur die observationes just seind. [g] *Margin*: waren. *Deleted*: sein gweßt. [h] *Added above the line*: sie. [i] *Deleted*: höher. [j] *Deleted*: stehen, als. [k] *Margin*: nit nur … als. [l] *Added above the line*: im himmel. *Deleted*: auch. [m] *Added above the line*: erhoben sein. [n] *Margin*: gegen ihrer höhen. [o] Seind … seye, *added later in a very small, barely legible script*.

come together, they form the parallax angle ACB 3°, the proportion of whose sine to AB is the same as the proportion of sine ABC to AC, and thereby the length of line AC is measured no differently than if it had measured the same line from A to C in the air with a measuring rod or meter stick; in the same way, that is, according to the demonstration of the following figure on page 20, Copernicus[168] | measured the distance of the Moon from the Earth. However, 41
he found the angle BAC or DAC at 80°55′ and observed the angle DBC at 81°55′.[169] By this means, he flew up to the Moon itself and measured the angle or parallax ACB, since the two lines AC and BC came together at the Moon (at the hour of his observation), and found that it was 1° or 60′.[170] As further borne out by geometry and arithmetic, sine 1747 of parallax angle ACB 1° measures the same proportion to AB, or one earth radius, as sine 99006 of angle ABC to AC. In this way, Copernicus measured line AC, that is, how far the Moon stood (at that time) from the center of the Earth, and found that the distance was 56 2/3 earth radii.[171]

In the very same way, as also mentioned above in Chap. 5, geometry provides infallible demonstrations to reckon the distance of comets and shows how to determine their parallaxes and distinguish them; for example, if a comet were closer to the Earth than the Moon, the measure of parallax (in the previous scheme, ACB) would have to be that much greater. If it is farther away, however, the parallax (AMB) must thus be that much smaller. From this, it follows irrefutably (**for geometry and arithmetic will never let you go wrong**) that since the parallax of every comet that has been properly observed up to now has been much smaller than that of the Moon, and in the greater part [of comets] no parallax has been discerned at all (as already proven above in Chap. 6 in the case of this final comet of 1618),[172] so it must certainly follow that not only are they situated higher above the Earth than the Moon, but they are elevated so high in the heavens that half the Earth's thickness can no longer bear a sensible proportion to their height, and it is to be reckoned as naught or nothing.

On this point, Bartholomaeus Keckermann in particular declares his defense vigorously, along with many others, in the *Systema physicum*[173] and the *Dis-*

168 *Margin: lib, 4, cap. 16 et 17*. Copernicus (1543, 118^{r}–120^{r}); cf. the image on p. 119^{v}.

169 Copernicus (1543, 119^{v}).

170 Copernicus (1543, 119^{v}).

171 Copernicus (1543, 119^{v}).

172 See pp. 24–25 above.

173 Keckermann (1623). The first edition was published in 1610. A first draft of this presentation of Keckermann's criticism is present on p. 2^{v} in the slips that represent the intermediary state.

Physicis, Disput. extraord. de Cometis, et adiuncta διασκέψει. Da er sihet, das
parallaxeon observatio ihme vil zu hell ist, und gar zu scharf under die augen
scheine, damit[a] sein gefaßte (doch wider *Aristotelis* protestieren) opinion gar
verdunckelt wirt:[b] daneben er auch in seinem Gewißen überzeugt ist, das,
was man von den Cometen grundtliches und gewises will und kan wißen, sol-
ches alles stehe, und müße gegründet sein, eintzig und allein auf die Paralla-
xes: Darumb so understeht er sich *per fas et nefas* die *parallaxes* nit aufkom-
42 men zu laßen, sucht allerlay ein und ausreden, ob | er möchte erhalten, alles
was von *parallaxibus* zu observieren fürgebracht wirt, verdächtig zu machen.
Dann, gibt er für, es seye nicht nur den *Instrumentis*, damit man observiert,
wie auch den *mediis, per quae*, dem Lufft, dem Element des fewrs, und him-
mel, durch welche das Gesicht die Stern oder Cometen faßet, sonder auch
den *Mathematicis* selbsten, nit zu trawen. Aber, wiewol er in andern, son-
derlich in *Logicis* ein fürtreffenlicher *Philosophus* war (ohn das ers in *Theo-
logia*, wie meniglichen bewußt, übel mißbraucht) so verräthet er in *Mathe-
matis* sein groben unverstand und unwißenhait, oder wans aufs höchst solte
komen, sein genaigt sinn und gemüt, alles zu verkehren, damit er sein einge-
bildte meinung möchte erhalten. Seine nichtwoerde Einwürff hie zu erzeh-
len, ist on not, sie gehören an ein ander ort. Jedoch nur ettlichs anzuregen:
Er solte doch in allem, was er anklagt, ein underschaid wißen zu halten. Es
ist ja nit ein ieder ein *Mathematicus*, welcher sich deßen im Argwon hatt,
als zum Exempel (*licet Exempla sint odiosa*) Er Keckerman selbs. Doch solt
er auch wißen, das under den *Mathematicis*, wie in allen andern künsten,
ein underschid ist. Alle *scientiae*, und jede in sonderhait, sind *infinitae*, und
unausforschlich, keiner kans volkommen faßen, *animus et intellectus hominis
est finitus*, darumb thailts Gott, wie alle seine Gaben, also *Mathematum sci-
entiam* oder vil mehr *scientias*, under den Menschen aus nach gewißer Maß.
Mancher ist ein fürtreffenlicher *Mathematicus* oder *Astronomus*, und waißt in
Theoria, oder wie man sagt, daheim hinderm Ofen, alles *solidè*, recht, wol, und
ohne fähl zu demonstrieren: wann er aber *ad praxin* kompt, und will rechnen,

---

[a] *Deleted*: ist und. *A barely legible addition in the margin*: ist, und gar … damit. [b] *Margin*: wirt.

---

174 Keckermann (1606). This work was reprinted in 1617; we refer to it by the first edition. The Διάσκεψις *de observationibus cometarum per instrumenta Astronomica, atque adeo etiam eius, qui apparuit Anno praecedenti 1604. Octobri, Novembri et Decembri mensibus*, occupies pp. 377–417. This intervention follows a series of *Theoremata exegetica De Cometis, in genere; Et in specie, de tribus illis mirabilibus Facibus, que Anno 1572. et denique Anno praeterito 1604. apparuerunt. Quae* […] *in Gymnasio Dantiscano Praeside B. Keckermanno* […] *Extraordinariae Disputationis examini subiecit Petrus Crugerus*, pp. 344–376.

putationibus Physicis, Disput[*atio*]. *extraord*[*inaria*]. *de Cometis*, and the joined διάσκεψις.[174] There, he views the observation of parallax as far too bright for him and shining all too sharply before his eyes, such that his established opinion (though against the protest of Aristotle)[175] is obscured entirely; in his conscience, he is even convinced that anything one can and wishes to know about comets certainly and soundly, all depends, and must be founded, entirely and exclusively upon parallax. For this reason, he does not concede, for right or wrong, to allowing for the acceptance of parallax, but seeks every sort of exception and excuse to see whether | he may maintain that all that is brought for- 42
ward from the observation of parallax is rendered suspect. For he pretends that it is not only the instruments with which one observes, as well as the means through which one grasps the stars or comets by sight, including the air, the element of fire, and the heavens, but also mathematicians themselves that are not to be trusted. Yet, as much as he was[176] an excellent philosopher in other subjects, especially logic (with the exception that he applied it poorly in theology, as proven by many), he thus reveals in mathematics his blatant lack of understanding and ignorance, or, if it should come right down to it, his own inclination and will to turn everything around in such a way that he might maintain his own imagined opinion. There is no need to recount his hollow objections here, they belong elsewhere. However, there is something worth suggesting: he should, in everything that he condemns, still know how to maintain one distinction. For not everyone is a mathematician who considers his own art with suspicion, such as, for example (if we may allow for offensive examples), Keckermann himself. He should also know, however, that among the mathematicians, as in all other arts, there is a difference. All sciences [*scientiae*], and every one in particular, are infinite and inscrutable, and no one can grasp any of them fully, for the mind and intellect of man is finite, which is why God, as with all his gifts, distributes the science, or rather sciences, of mathematics among men according to a certain measure. There are some excellent mathematicians and astronomers, and in theory or, as they say, at home behind the oven,[177] they know how to demonstrate everything firmly, fairly, well and

175 See pp. 36–37 above and notes 148–150. On Keckermann's criticism, see Introduction, Chapter 6 and Gindhart (2006, 234–243); on Maestlin's reply, see Introduction, Chapter 7.

176 Keckermann died in 1609.

177 On this expression and other equivalent ones ("ein mathematicus underm Dach"; "im warmen Stüblein undt unter dem Dach"), indicative of a certain suspicion about pure theoretical astronomy, see Christoph Rothmann's letter of 11 December 1585 to the Landgrave Wilhelm IV of Hesse-Kassel (Granada, Mosley, and Jardine (2014, 240)) and the Landgrave's letter to Brahe of 14 April 1586 in Brahe (1596, 49).

meßen,[a] oder observieren, so fählt es im am handtgriff bisweilen nit wenig, wie droben, cap. 7, am Johanne Vögelin, Andrea *Nolthio*, und andern gezaigt worden. Und ob schon dis also ist: so folgt doch gar übel, wie Keckermann wol waißt, *à parte ad totum* argumentieren wöllen, das drumb alle (darunder er auch des Tychonis nit verschonet)[b] müßten gefählt haben. Tycho Brahe, und ein jeder geübter *Artifex*, wüßten aus gnugsamer Experientz, wie die *Instrumenta* anzurichten, desgleichen, was, wie, und wan zu observieren,[c] und jedes also anzustellen seye,[d] damit aller verdacht und argwohn einiges Ihrthumb wol verhüttet werde,[e] dann sie seind nit Kinder, wie Keckermann sie gern schmitzen wolt. Dis und dergleichen solte Er zuvor bedacht und examiniert haben,[f] ehe er alles über ein hauffen hatte geworffen. Aber understehe er sich, wie er wölle, *Parallaxin* wirt weder Er, noch kein anderer umbstoßen. *Geometria* hatt sie zu starck gegründet.

43 Nach dem also die Cometen aus dem Elementarischen Lufft erledigt, und in ire[g] ordenliche Stell wider heimgebracht seind, so folgt jetzt, das was auf dise *falsam hypothesin* und mirben[h] grund ist aufgebawen, sampt vilen scharpsinnigen disputationen, auch fallen müßten, under denen sind: das die Cometen aus den von der Erden auffsteigenden Dämpfen herkommen; das sie[i] vom Element des fewrs droben entzündet werden; dann sie müßten je aus einer andern materi iren Uhrsprung haben,[j] wie auch ir liecht oder schein anderstwo her empfangen: ob die aufsteigende Dämpf so hoch kommen könden, welcher mainung die *Optica, in doctrina de Crepusculis* widerspricht? Item ob das Element des fewrs, und obere thail des Luffts, von dem Himmel, *virtute primi Mobilis*, wie die andere *Sphaerae coelestes* (*secundum usitatas hypotheses* zu reden) umb die Erden täglich herumb geführt werden? Ja es zweiflen auch ettliche fürtreffenliche *Optici*, ob ein *Sphaera Ignis* droben seye? Aber hiervon zu dißerieren, gehört auch anderstwo hin.

Das aber *Keckermannus* fürgibt, besorgt, und sich deßen gleichsam vor Gott und der welt beklagt: wann man von den Cometen nit lehren solte, wie Aristoteles gthon, so würde die gantze *Physica scientia*, und mit ir alle andere *disciplinae*, geschwächt, aller dings vernichtet werden, und zu scheittern gehen, darumb rüfft er alle, so *philosophiam* lehren, umb hilf und beystand an, das sie müglichsts fleis helfen wehren, damit die Cometen nit in den himmel nistern etc.

---

[a] *Added above the line*: meßen. [b] Parenthesis added in the margin. [c] *Deleted*: seye. [d] *Added under the line*: seye. [e] *Verdacht ... werde* appeared in the first version in the first line of the following page. However, Maestlin decided to add *dann sie seind ... gegründet* at the bottom of p. 42. [f] *Deleted*: (doch wan ers geköndt hette). [g] *Added above the line*: ire. [h] mürbe. [i] *Added above the line*: das sie. [j] *Margin:* iren Uhrsprung haben.

without fail, but when it comes to practice and they wish to reckon, measure, or observe, they do not manage at times to find even a fraction of a firm grasp, as was shown above in Chap. 7 with Johannes Vögelin, Andreas Nolthius, and others.[178] And although this is the case, it follows quite poorly, as Keckermann well knows, that all of them, by arguing from the part to the whole (even Tycho was not spared among them), should have failed. Tycho Brahe and every skilled *artifex* would know from enough experience how to arrange the instruments, among other things, as well as how, what, and when to observe with them, and so to set up everything in such a way as to avoid the slightest hint or suspicion of any error, for they are not children, as Keckermann would like to scorn them. This and other such things he should have first considered and examined before he messed the whole matter up. Yet let him stand as he may wish, neither he nor anyone else will overturn parallax. Geometry has founded it too firmly.

Now that comets have been cleared from the elementary air and brought 43
back home again to their proper place, it thus follows that what was based on this false hypothesis and feeble foundation, together with many clever disputations, must also fall, among them being: that comets originate from exhalations arising from the Earth; that they are ignited by the element of fire up above; for they would then have to draw their origin from another material, just as they have received their light or luster from another source; could the rising vapors reach that high, which opinion does optics contradict by the doctrine of twilight [*doctrina de crepusculis*]? Further, is the element of fire, as well as the upper part of the air, driven around the Earth daily by the heavens,[179] through the power of the *primum mobile*, like the other celestial spheres (speaking according to the standard hypotheses)? To be sure, many distinguished opticians doubt whether there is even a sphere of fire above us.[180] However, there is a better place for disputing such a subject.

Yet what Keckermann submits, worries about, and, as it were, laments before God and the world is that, should we no longer teach about comets in the same way as Aristotle did, natural philosophy as a whole, and with it all other disciplines, would be weakened, no doubt destroyed, and fall to pieces, which is why he calls on all who teach philosophy to help as much as possible through their service and support to negate that comets nest in the heavens, and so forth.[181]

---

178 See pp. 28–29 above.

179 Cf. p. 40 and note 166.

180 See Pena (1557, f. bb ij): "Ergo docet Optice somnium esse, quidquid de ignis sphaera Physici disputant" (Preface *De usu Optices*).

181 Cf. *Diaskepsis*, in Keckermann (1606, 380): "Non potest sane salvum nobis manere amplius

Antwort: *Mi Keckermanne*, weniger ists nicht, es würde ettlichs, so bisher für gewis und wahr gehalten worden,[a] selbiges aber *in rei veritate* sich anderst befindet, darnider ligen. Aber[b] es ist ein unnotwendige fürsorg, das wegen deßen solte der himmel einfallen, und die gantze *Physicam scientiam* erschlagen. Sonder hiemit, wann man dis, was nit wahr ist, ab sich uffs und aus mustert, hergegen aber lehret, was mit grund der warhait erwisen wirt, geschicht keiner
44 *Scientiae* einiger Abbruch, sonder | sie werden desto klärer illustriert, verbeßert und was noch verborgen ist, ans tagliecht gebracht, darfür man auch Gott dancken soll.[c] Oder, will Keckermann, dem *Aristoteli* (doch wie ettlich mal gemeldt, wider sein willen) zu gefallen, vil lieber falsch und unrecht, als Gott zu ehren, und[d] der lieben Warhait zu stewr,[e] recht lehren? Ein hocherleuchter *Philosophus* ist *Aristoteles* geweßt, ab deßen *divino ingenio*, und von Gott ime[f] verlihenem hohen Verstand sich billich meniglich verwundert, niemandts aber bisher ime gleich gethon darumb auch was er gelehrt, darvon soll niemand abweichen, er seye dann vil beßer gegründet:[g] jedoch beklagt er sich bisweilen, das es ime an einem und andern gemanglet habe, wie alberait hieoben in der Cometen materi geschehen, und Er auch, da er von *Galaxia*, oder *circulo Lacteo*, der weißen Straß am himmel, so die Hör, oder S. Jacobstraß genennet wirt, handelt, nit in Abred ist, da er sagt, *lib*. 1. *cap*. 8. *Meteor*. Dieweil wir die Ursachen (Ursprung und Zustand) der Cometen also annemmen und gutthaissen, wie sie füglich (ἀτρίως) fürgebracht worden: Also soll man[h] darfür halten, es

---

[a] *Deleted*: fallen. [b] *Added at a later date*: darnider ligen. Aber. [c] *Margin*: darfür … soll. [d] *Margin*: Gott zu ehren, und. [e] *Deleted*: und Ehren. [f] *Added above the line*: ime. [g] *Margin*: darumb … gegründet. [h] *Deleted*: auch.

---

divinae istius scientiae, quae de rerum natura tractat, Corpus ac Systema, si doctrina de caelo, de igne, de aere, de meteoris nulla amplius certa sit: est enim ista prima Physicae specialis pars, quae si cadat, continuo secum trahet ruinam reliquarum omnium in ista scientia doctrinarum, nempe de corpore mixto, de metallis, de plantis, de animalibus […] unde verendum fuerit, ne pro Systemate scientiae naturalis, opinionum variarum ac dubiarum Chaos in scholis simus habituri"; 381: "Quocirca vehementer optem, istam de fumis ad ipsum usque aethera evectis, cumque purissimo illo corpore caelesti confusis, ibidemque accensis, opinionem vel non audiri plane, vel ampliari saltem in Philosophorum scholis cum illa solenni formula: NON LIQUET; neque enim tam facile restaurantur bonae disciplinae, quam convelluntur". Cf. also the conclusion, pp. 414 f. and Introduction, p. 30.

182 Cf. p. 25 in the rough copy: "Das man aber fürgibt und besorgt, die *physica* werde hiedurch allerdings fallen, etc., ist ein onnothwendige fürsorg, das wegen dises stuks in der *physica* der himmel solte einfallen, und die gantze *physicam scientiam* erschlagen, sonder beides desto mehr erleuttern und was darinnen noch mangelhaft oder dunkel, kompt desto heller

Answer: My Keckermann, there would nevertheless be some things, still seen as certain and true, that in reality prove different and remain poorly rendered. However, it is a needless concern that, on account of this, the sky should fall and all of natural philosophy lie in ruins. Rather, when one thus refutes exclusively what is not true and, in its place, teaches what is proven on the basis of truth, none of the sciences are doomed to destruction, but | are illuminated and 44
improved all the more plainly, and what still lies hidden is brought to light, for which we should thank God as well.[182] Or does Keckermann prefer to oblige Aristotle (even, as mentioned above many times, against his own will),[183] to teach far more false and wrong, than to honor God, defend the dear truth, and teach properly? Aristotle was a highly enlightened philosopher who rightly astonishes many with his divine intellect and great God-given genius, and no one has ever measured up to him, which is also why no one should deviate from anything he taught, since he founded it far better;[184] yet he laments at times that he has fallen short on one subject or another, as already happened above on the material of comets, and he does not deny it when he deals with the galaxy or Milky Way, or, as is heard, the white street in the sky, also known as St. Jacob's Street, when he says in Book 1, Chap. 8 of the *Meteorologica*: "Since we thus accept and approve the causes (origin and state) of comets as they are put forward as fitting (ἀτρίως),[185] so should we assume that the very same possi-

---

an tag". This passage follows the criticism of contemporary Aristotelians on the same page (p. 39 in the clean copy). For a similar passage in the manuscript treatise on the nova of 1604, see Granada (2014, 114, 121): "It is not that anyone should believe that in this way the old arts and sciences are to be destroyed, as if proceeding on this path a new one is finally being sought by us. But the true and ancient sciences are not being unsettled, much less are they being overturned; however, the erroneous opinions in them, having been defended as true up until now, are pulled out and removed, and in their place is planted the indubitable Truth. Moreover, by the common chain of Truth all doctrines are elaborated and accepted, and sciences, the more eminent as well as the more skilful, are connected, for all of them are established and come from God, who is the truth itself". For a similar statement following the telescopic observations of the Moon and the Sun, see p. 55 below.

183 Maestlin refers again to the 'protestation' of Aristotle in *Meteorology*, I, 7. See pp. 36–37 above.

184 Aristotle. The following section on the Milky Way (pp. 44–46) appears for the first time in the slips of paper representing the intermediary state (fascicle 8b, pp. 7–8).

185 Maestlin translates *Meteorology*, I, 8, 346b 1–3: "Hence if we accept the cause assigned for the appearance of comets as *plausible* we must assume that the same kind of thing holds good for the Milky Way". The Greek text actually established presents μετρίως instead of ἀτρίως.

habe mit dem *circulo Lacteo* eben auch dise gelegenhait, etc. Aristoteles erklärt sich am selbigen ort, wiewol die vorige Alte *philosophi* nicht aller dings ainig, so stimmen sie doch samentlich in disem überein, das *Galaxia* stehe droben im Himmel: Er aber halte es darfür, das die Cometen und *Galaxia* haben ainerley Uhrsprung, substantz und wesen, nämlich von den aus der Erden aufriechenden Dämpfen: Sie seyen beede im obern thail des Luffts; daselbsten werden sie gleicher gestalt von krafft des fewrs angezündet: Allein mit disem Underschaid, das der Cometen stromen oder schein allein[a] von einem Stern oder haupt ausgehen, in *Galaxia* aber oder in der Jacobs Straß erstrecke sich der Schein von vilen sternen, und figurieren einen gantzen *Circulum magnum*: Und, der Comet verlesche wider, die Jacob Straß oder *Circulus Lacteus* aber bleibe stetigs. Dann die Dämpf, wie sie aufsteigen, ersetzen als bald wider, alles was sich an jedem ort verzehrt, hiemit werde aller | Abgang wider ausgebießet. Wan aber dieselbig materi zu haüffig aufsteiget, so samle sie sich, und werde ein Comet daraus, darumb bleibe die Jacobs straß stetigs in einer form und große, die Cometen aber erscheinen selten.

Hie sagt *Aristoteles* abermal nicht, das er sein fürbringen von den Cometen mit sattem grund habe demonstriert und bewisen, sonder er hab beschaidenlich, *moderatè et probabiliter*, so gutt es sein mögen, darvon geredt. Ey wie hette *doctrina parallaxeon* alda[b] so wol gethon! Were sie dazumal erfunden gweßt, Aristoteles hette freylich weit anderst von Cometen geschriben, und die Jacobs Straß nicht auf ein so bawfälligen grund gesetzet.

Das aber dise Jacobs Straß nit im Lufft, sonder im himmel am firmament seye, wirt bewisen: Erstlich, dieweil an allen deren orten kein *parallaxis* gespirt wirt. Darnach dieweil in so vil hundert Jaren sie sich nit verendert hatt, sonder wie sie von *Ptolemaeo lib.* 8. *cap.* 1 *Almag.* und von andern noch eltern *Philosophis* ist aufgezaichnet worden, also[c] ist sie an größe, und bey ainerlay fix[d]

---

[a] *Added above the line*: allein. [b] *Added above the line*: alda. [c] *Margin*: also. [d] *Margin*: fix.

---

186 Cf. Aristotle (1984, I, 8, 345b 35–346a 10): "We must suppose that what happens [in the Milky Way] is the same as in the case of the comets when the matter does not form independently but is formed by one of the fixed stars or the planets. Then these stars appear as comets, because matter of this kind follows their course. In the same way, a certain kind of matter follows the sun, and we explain the halo as a reflection from it when the air is of the right constitution. Now we must assume that what happens in the case of the stars severally happens in the case of the whole of the heavens and all the upper motion. For it is natural to suppose that, if the motion of a single star excites a flame, that of all the stars should have a similar result, and especially in that region in which the stars are biggest and more numerous and nearest to one another". Maestlin introduces the issue of the Milky Way (as well as later, pp. 46 ff. the habitation of the torrid zone)

bility applies to the Milky Way, and so forth."[186] Aristotle explains in the same place that, although earlier philosophers of old certainly did not agree entirely, they did concur completely that the Milky Way may be found on high in the heavens;[187] however, he would hold that comets and the Milky Way share the same origin, state of being, and substance, namely from the exhalations arising from the Earth, that they are both in the upper part of the air, where they are ignited by the force of fire in the same way; with the one difference being that the stream or glow of comets emanates only from one star or head, while the light from the Milky Way or St. Jacob's Street extends from many stars that fully form a great circle; and the comet extinguishes again, while St. Jacob's Street or the Milky Way remains constant. For as the exhalations rise, they occupy at once the place of everything consumed there before it, and thus clear the way again for all | outflow. Yet if the same material should 45
arise too often, it may gather together and become a comet, which is why St. Jacob's Street forever maintains the same form and size, while comets seldom appear.[188]

Here, once again Aristotle does not say that he has demonstrated and proven his assertion about comets with absolute reason, but that he has spoken of them prudently, soberly, and probably,[189] as well as may be. Oh, how the doctrine of parallax would have served [him] so well here! Had it been invented at that time, Aristotle would have certainly written far differently about comets and not situated St. Jacob's Street on such dilapidated ground.

That St. Jacob's Street is not in the air, but in the firmament in the heavens shall now be proven: first, because at no point in the Milky Way is parallax ever perceived; second, because it has not changed in so many hundreds of years, but, just as it was recorded by Ptolemy in Book 8, Chap. 1 of the *Almagest*[190] and by other, more ancient philosophers, it has remained the same in size

---

in order to weaken Aristotle's authority in natural philosophy and cometary theory, since the Aristotelian explanation of the Milky Way as a sublunary, permanent and identical, phenomenon (as Maestlin later describes it) had been rejected since antiquity. For the history of the problem, see Jaki (1973) and Telesio (2012, Introduction).

187 Aristotle (1984, I 7, 344b 12–18, 345b 9).

188 Aristotle (1984, I, 8, 346b 3–10): "For the tail which in the former case [a comet] is an affection of a single star here forms in the same way in relation to a whole circle. So, if we are to define the Milky Way, we may call it the tail attaching to the greatest circle, and due to the matter secreted. This, as we said before, explains why there are few comets and why they appear rarely; it is because at each revolution of the heavens this matter has always been and is always being separated off and gathered into this region".

189 Cf. p. 44 and note 185.

190 Ptolemy (1984, VIII, 2, pp. 400–404).

Sternen verbliben. Deswegen haben alle *Philosophi* nun mehr ein lange Zeit des *Aristotelis* meinung ausgesetzt. Dises erkhennet und bekhennet Keckermann auch selbs, in *Disputationibus Physicis, Disput.* 8. *in fine.* Doch ein blawen dunst für die Augen zu machen, setzt er hinzu: **Es geduncke in, es were *Aristotelis* meinung mit *Plinio, lib.* 18. *cap.* 29. und andern *Philosophis* wol zu verainigen,[a] wie er dann solches durch gelegenhait im disputieren wölle anzaigen.** Ich aber hab dise sein vergleichung noch nit gesehen; kan auch nit erdencken, wie doch dise zwo so widerwertige *Propositiones*, und Reden: **Die Jacobs Straß steht nit droben im firmament, sonder hieunden im Lufft**; und, **die Jacobs Straß steht nit im Lufft, sonder[b] zu öberst im aller höchsten himel im hohen firmament**, mit einander zu vergleichen weren, wie dann einige Vergleichung nimmermehr erfolgen kan.

46 Weil er dann alhie muß von *Aristotelis* meinung ab|tretten, und die Jacobs Straß in[c] dem himmel laßen, so solte er den Cometen, als überwisen, den himmel[d] auch nit mißgünnen, sonder vil mehr gedencken, das von *Aristotelis* zeitten noch anders mehr ist erfunden worden, welches im nie in sinn kommen, oder darvon er weit[e] andere gedancken gehabt, und ist doch sein *Philosophia* deswegen nicht zugrund gangen, sonder nur desto mehr verbeßert worden.

Als zum Exempel, zu *Aristotelis* Zeiten ist man in gemein beredt geweßt, das in *Zona Torrida*, zwischen beeden *circulis Tropicis*, in denen orten der Erden, da die Sonn jährlich zu gewißen Zeit ob dem Haupt her gehet, wegen unleidenlicher großen Sonnen hitz, kein Mensch wohnen könde, dann so die Sonn zu Sommers zeit so hais bey uns einbrennet, da sie doch von unserm Zenith oder Hauptpuncten noch zimlich weit stehet, wie warm wirt sie einhaißen, da sie gar ob dem haupt her gehet, hierauf gründet *Aristoteles* dis, da er *lib.* 2, *cap.* 5. *Meteor.* schreibt, das nicht allein an selben orten, sonder auch ehe man zu dem *Tropico* kompt, niemandt wohnen könde.

Das aber selbige Ort nit also öd, und ohn Einwohner seyen, hatt man in folgenden Zeitten gnug erfahren. *Ptolemaeus* bey anderhalb hundert Jar nach *Christi* geburt, beschreibt *lib.* 4. *Geographiae, Tabula* 3. et 4. *Aphricae*, und *lib.* 5. *Tab.* 6. 9. 10. 11. 12. *Asiae*, ein zimliche anzal Ort und Landtschafften, welche innerhalb der *Zonae Torridae* bewohnet werden, deren zum thail über den *Aequinoctial* hinüber gegen Mittag bis in 16. grad ligen, das seind schier dritthalb hundert, und über den *Tropicum Cancri* oder der *Zonae Torridae* anfang, bey 600. teutscher meilen.

[a] *Deleted*: wol zu vergleichen. [b] Margin: nit im Lufft, sonder. [c] *Added above the line*: in. [d] *Margin*: den himmel. [e] *Added above the line*: weit.

191 Keckermann (1606, [*De Meteoris apparentibus disputatio octava*] 100).

and in relation to the same fixed stars. For this reason, all philosophers have abandoned Aristotle's opinion for a long time now. Even Keckermann himself acknowledges and admits this near the end of Disputation 8 in the *Disputationibus physicis*.[191] Yet to pass a pure fancy before our eyes, he adds that **he believes the opinion of Aristotle could very well be aligned with those of Pliny in Book 18, Chap. 29, and other philosophers, as he would then wish to show through the opportunity of disputation.**[192] However, I have still not seen this comparison of his, nor can I imagine how two such contradictory propositions and lines of discourse—**St. Jacob's Street is not located up above in the firmament, but here below in the air**, and **St. Jacob's Street is not located in the air, but in the uppermost part of the highest heaven in the outer firmament**—could be compared with one another, for any such form of comparison can never be made.

Since he must then step | away from the opinion of Aristotle at this point and 46
leave St. Jacob's Street in the sky, so he should also not hold the heavens against comets, as shown above, but consider more carefully that from the time of Aristotle many more things have been found that never occurred to him or that he thought about far differently, and this is thus why his philosophy did not break down but was improved all the more.

As an example, it was common currency in the times of Aristotle that, in the torrid zone between the two tropic circles, in those places of the Earth where the Sun passes overhead at a certain time every year, no one could live due to the insufferably strong heat of the Sun, for the Sun burns us so acutely in the summer even when it is so far from our zenith or highest point that Aristotle establishes by how warmly it will heat the earth where it passes directly overhead that, as he states in Book 2, Chap. 5 of the *Meteorologica*,[193] not only could nobody live in the above places, but even before one reaches each tropic.

That the same places are not so barren or bereft of inhabitants was sufficiently understood, however, in later times. Ptolemy, around 150 AD, describes in the *Geographia*, Book 4, Tables 3 and 4 of Africa, and Book 5, Tables 6, 9, 10, 11, and 12 of Asia,[194] quite a number of landscapes and locations within the torrid zone that are inhabited, some of which lie beyond the equinoctial line by as much as 16° below the equator, that is, nearly 250 [German miles], and beyond the Tropic of Cancer or the start of the torrid zone by 600 German miles.

192 *Ibidem*: "Verum Aristotelis sententia cum sententia Plinii et aliorum Philosophorum conciliari posse videtur, ut per disputandi occasionem indicabimus".

193 Aristotle (1984, 362b 21–27). For a neutral presentation of this issue, see Maestlin (1610, 158); for an excellent collection of studies on the history of the matter, see Randles (2000).

194 Tables 6, 7, and 9 of Asia correspond to book VI, and tables 10 and 11 to book VII.

Was soll man von unsern Zeitten sagen? Seidher von *Anno* 1492. (sind noch nit 130. Jar) hatt man gantz *Americam* erkundigt, welche so vil hundert jar unbekandt gweßt, darumb mans jetzt, die Newe Welt, haißt, deren ein großer thail sich durch *Zonam torridam* hindurch[a] erstrecket, und doch wol bewohnet
47 wirt. Hiemit hatt | man noch beßer erlernet, das man nit alein[b] durch *Zonam Torridam* wegen großer hitz, wol und sicher kommen kan, dann es ist deren, so aus Europa hindurch geschiffet, keiner von der Sonnen gebraten oder ausgedorret: sonder man hatt auch erlernet, das man umb die gantze Welt, oder gantzen Erdkreis herumb schiffen kan. Dis alles ob schon des *Aristotelis* fürgeben ungleich und zuwider lauttet, so wirt doch weder *Geographiae*, noch *Matheseos* noch auch *Philosophiae studium* geschmählert, sonder desto mehr erweittert.

Allein ist sich zu verwundern, das *Aristoteles* an gemeldtem ort, *lib.* 2. *cap.* 5 *Meteor.* zum andern mal schreibt, man könde ettliche ort auf der Erden, wegen grausamer kältin (als da seind beede *Zonae frigidae sub polis*) ettliche wegen großer hitz (als *Zona torrida*, zwischen beeden *Tropicis*) nit bewohnen, und setzt, wie gemeldt, hinzu, **das auch ehe man zum *Tropico* kompt, niemandt sich aufhalten könde**: hatt dann *Aristoteles* nit gewüßt, das es mit *Zona torrida* anderst beschaffen gweßt?

Als: Meröe war je und je ein weitberhümbte Stat, sie ligt in einer Insel im Nilo, auch Meroë genandt. Den Namen hatt sie empfangen bey 200. Jar vor *Aristotelis* Zeitten, von *Cambyse* König in Persia, des Königs *Cyri magni*, welcher das Jüdische Volck aus der Babylonischen Gefäncknus hatt frey wider laßen heim ziehen, Son. Diser Cambyses,[c] als er Egypten mit Krieg bezwungen, und *Aethiopiam* oder die Mohren auch anzugreiffen vorhabens war, ist er bis in Meröe komen, selbige also genennet, wegen seiner Schwester, oder wie ettliche wöllen, Gemahl, welche daselbsten gestorben, und begraben worden. *Elevatio poli* diser Stat ist 16. gr. 26. min.[d] Das ist, bey acht halb grad über den *Tropicum Cancri* hinein werts, welches ist über 110. teutscher meilen. Sie war die Haupstatt in *Aethiopia*,[e] welches als ein herrlich und mächtig land, auch über *Syriam* (und also über vil Länder weit und brait) geherschet hatt. Daher die Mohren nit nur vil fältig mit den Egyptern, als mit iren nahen Nachbaurn, schwere Krieg

---

[a] *Added under the line*: hindurch. [b] *Added above the line*: nit alein. [c] *Margin*: Herodotus lib. 3. Strabo lib. 17. [d] *Margin*: *Ptolem. Tab. 4 Aphricae.* [e] *Margin*: *Diodor. Sic. lib. 1. et 3. Plin. lib. 2. cap. 73. Lib. 6. c. 29. Lib. 5. cap. 9.*

---

195 See Aristotle (1984, II, 5, 362b 7–9): "Beyond the tropics no one can live: for there the shade would not fall to the north, whereas the earth is known to be uninhabitable before the shadow disappears or it is thrown to the south". For the first mention, see note 193.

What should one tell from our own times? Since the year 1492 (that is, not quite 130 years), all of America has been explored after remaining unknown for so many hundreds of years, which is why we now call it the New World, a large part of which extends well into the torrid zone and is still widely inhabited. Hereby we came to learn | even more clearly that not only can one pass safely 47
and soundly through the torrid zone despite the severe heat, since no one who sailed through the area from Europe has ever been burned or dried up by it, but we also learned that one can sail around the whole world or entire globe of the earth. All of this, although it sounds unlike and even contrary to the position of Aristotle, shall not diminish the study of geography or mathematics or philosophy, but rather expand them all the more broadly.

It is only left to wonder why Aristotle states a second time in the above passage in Book 2, Chap. 5 of the *Meteorologica* that no one could inhabit several areas of the Earth due to their terrible cold (being as they are both frigid zones around the poles), and several others due to their great heat (as they belong to the torrid zone between the two tropics), and adds, as mentioned above, **that no one could live even before one reaches one of the tropics.**[195] Did Aristotle not know that things proved differently in the torrid zone?

For example, Meroe was at one time a widely renowned city that lies on an island in the Nile also known as Meroe, a name it received around 200 years before the time of Aristotle by Cambyses, King of Persia, son of King Cyrus the Great, who freed the Jewish people from the Babylonian captivity to return home [to Jerusalem]. When the same Cambyses[196] defeated Egypt in battle and was also planning to attack Ethiopia or the Moors, he came to Meroe and named it after his sister, or, as many prefer to say, his wife, who died and was buried there. The elevation of the pole of this city is 16°26′,[197] that is, about 8 $^1/_2$° below the Tropic of Cancer, which is more than 110 German miles. It was the capital city of Ethiopia,[198] which ruled as a magnificent and powerful land over Syria as well (and thus over many other lands far and wide). Hence, the Moors not only waged wide-scale war many times on the Egyptians, as well as

196 *Margin*: *Herodotus lib.* 3. *Strabo lib.* 17. See Herodotus (1920–1925, iii, 31–32); Strabo (1917–1932), xvii, i, 5, C 790.

197 *Margin*: *Ptolem. Tab. 4 Aphricae*. See Ptolemy (1584, 57) Accordingly, the latitude of Meroe contradicts Aristotle's concept of the uninhabitability of the torrid zone. Cf. p. 48 below for Meroe and Syene and the presumed navigation to Peru by the ancient Jews.

198 *Margin*: *Diodor. Sic. lib. 1. et 3. Plin. lib. 2. cap. 73. Lib. 6. c. 29. Lib. 5. cap. 9.* See Diodorus Siculus (1933–1967, I, 33, 1–3; III, 6, 3); Pliny (1938–1952), II, lxxi, 178; VI, xxxv, 182; V, ix–xi (a general description of Egypt).

48 geführt, sonder auch ins | Jüdische Land offter mal mit Hörskrafft eingefallen, wie sonderlich Serah der Mohrenkönig wider Assa den König Juda, mit tausent mal tausent man, und 300. wägen[a] (über 600. Jar vor *Aristotelis* Zeitten)[b] ausgezogen ist, und deswegen die Mohren, wiewol sie in *Zona torrida* wohnten, und vom Jüdischen Land weit entseßen, doch gleich so wol, als die nächste Syrier, Egypter und Araber, von den Propheten und sonsten in H. Schrifft,[c] für schädliche und abgesagte feind, verrufft werden. Und wie Volckreich sonderlich Meroe die Insel geweßt seye, zaigt *Plinius lib.* 6. *cap.* 29. das zur zeit, als die Mohren mächtig waren, (das ist, vor *Aristotelis* zeiten) sie hab dritthalb hundert tausent gerüster Mann ins feld geben könden, und doch darneben 4000. Mayster in allerlay handtwerken erhalten. Solte dann *Aristoteles* solches nit auch gewüßt haben?

Item, Syene gräntzet zwischen *Aethiopia* und Egypten.[d] Dise Stat und Insel ligt auch umbfangen mit den Nilo, nicht ausserhalb der *Zona torrida*, sonder gerad an derselben ersten Anfang, *sub ipso Tropico Cancri*.[e] Ezechiel der Prophet[f] gedenckt deren und Ihres Thurns, als eines namhafften Gräntzhaus zwischen beeden Königreichen, zur Zeit der Babilonischen Gefängnus mehr als 200. Jar vor *Aristotelis* Zeitten.

Der weise König Salomon hatt bey 700. Jar vor *Aristotelis* Zeitten seine Schiffarten gehn Ophir geschickt, welche in drey Jaren ein mal, mit seer reichen Schatz wider kamen.[g] Es sagt auch die Schrifft, Salomon habe seine Diener, mit Hirams Königs zu Tiro Dienern geschickt, dann selbige seyen gutte Schiffleut, und des Meers erfahren gweßt. Daraus abzunemmen, das dise Schiffarten schon, ohn zweifel vil Jar zuvor sind gebraucht worden. Ophir aber, wie die Gelerten gemeinlich darfür halten, ist Peru in America, dann alle umbständ stimmen hiemit überein, wie dann auch die Namen Ophir, und Peru, nach

---

[a]*Margin*: 2. *Chron.* 14 *vers.* 19. [b]*Margin*: (über ... Zeitten). [c]*Margin*: und sonsten in H. Schrifft. [d]*Margin*: *Plinius lib.* 2 *cap.* 73. [lxxv, 183]; *lib.* 5. *cap.* 9. [e]*Margin*: *Ptol. Tab.* 3. *Aphr. Strabo lib.* [f]*Margin*: *Ezech.* 29. *vers.* 10.; 30. *vers.* 6. [g]*Margin*: 1 *Reg.* 9. v. 27 et 10. v. 11. 22. 2. *Chron.* 8. v. 18. et 9. v.10. 21.

---

199 *Margin*: 2. *Chron.* 14 *vers.* 19 [properly 9].

200 Pliny (1938–1952), VI, xxxv, 186: "it used to furnish 250,000 armed men and 3000 artisans". Plinius (1601–1616), vol. II, 277 gives 4000.

201 *Margin*: *Plinius lib.* 2 *cap.* 73. [lxxv, 183]; *lib.* 5. *cap.* 9. [V, x, 59]: "Dicionis Aegyptiae esse incipit [Nilus] a fine Aethiopiae Syene".

their close neighbors, but also | invaded Jewish land, often with a massive 48
army, such as in the particular case of Zerah, King of the Moors, who set out against Asa, King of Judah, with thousands upon thousands of men and 300 chariots[199] (more than 600 years before the time of Aristotle); and this is why the Moors, even though they lived in the torrid zone far away from Jewish land, are still decried as much as the neighboring Arabs, Egyptians, and Syrians by prophets and others in Holy Scripture as a hostile and sworn enemy. And how populous the island of Meroe was in particular, Pliny indicates in Book 6, Chap. 29 that, at the peak of their power (that is, before the time of Aristotle), they could have sent 250,000 armed men into the field while maintaining 4,000 masters of every craft.[200] Should Aristotle not have known that as well?

In addition, Syene borders between Ethiopia and Egypt.[201] This city and island also lies surrounded by the Nile, not beyond the torrid zone but at the very beginning of it, just below the Tropic of Cancer.[202] Ezekiel the Prophet[203] commemorates them and their tower as a famous border marker between the two kingdoms at the time of the Babylonian captivity more than 200 years before the time of Aristotle.[204]

Around 700 years before the time of Aristotle, the wise King Solomon sent his ships to Ophir and they returned once every three years laden with treasures.[205] Scripture also says that Solomon sent his servants with those of King Hiram in Tyre,[206] for the latter were good shipmen that knew the sea well. From this, we may assume that those shipments had undoubtedly been brought many years previously. However, Ophir, as scholars generally suppose, is Peru in America, since all circumstances coincide with this, as well as the fact that the names Ophir and Peru, after the manner of the Hebrew language, yield one

202 *Margin*: *Ptol. Tab.* 3. *Aphr. Strabo lib.* Cf. Ptolemy (1584, 54), where its latitude is given as 23° 50′; Strabo (1917–1932), II, v, 7, C 114.

203 *Margin*: *Ezech.* 29. *vers.* 10.; 30. *vers.* 6. Cf. Ezekiel 29: 10: "Behold, therefore I am against thee, and against thy rivers, and I will make the land of Egypt utterly waste and desolate, from the tower of Syene even unto the border of Ethiopia" (King James version).

204 Ezekiel was born ca. 622 BC.

205 *Margin*: 1 *Reg.* 9. v. 27 et 10. v. 11. 22. 2. *Chron.* 8. v. 18. et 9. v.10. 21.

206 See the above references in the previous note.

art hebraischer Sprach, *per metathesin literarum*, ain[a] wort geben. In disen Schiffarten haben sie, laut aller Landt und Meer tafeln, auch müßen durch *Zonam Torridam* fahren, dann sie haben ins Rote Meer eingesetzt, und seind doch unverletzt hindurch und wider herdurch, mit reicher waar heimkommen. Bleibt also darbey, das *Aristoteles* zu seiner Zeit,[b] wegen der grewlichen hitz in *zona torrida*, ist unrecht beredt gweßt. Jedoch ist dis alles wol zu gedulden, und gibt seiner *Philosophiae*, und was er sonsten wol gelehret hatt, aufs wenigst keinen Nachthail.

49 [c] Was sonsten unzahlbare sachen und künsten nach *Aristotelis* zeiten seind erfunden worden, ist nit auszusprechen. Innerhalb zwayhundert nächst verschinen Jaren (weitter hindersich nit zu rechnen) wie vil ist auf die Ban kommen? Under welchen sollen billich gezehlt werden die zum thail nutzliche, zum thail aber die hochschädliche pichsen, welche zwar zur defension und zum schirm[d] manchem wol bekommen, sie werden aber zur offension, laider, zur belaidigung und schädigung[e] vil mehr gebraucht. Wie vil hundert tausent mal tausent menschen haben, seid her sie aufkomen, elendiglich und unschuldiglich ir Leben dardurch verlohren? Hingegen ist die hochlöbliche und nutzliche Buchtrucker kunst erfunden worden. Mit was großer mühe und arbeit (des unkostens zu geschweigen) hatt man zu vor die Bücher müßen abschreiben, da sie doch kaum dem hundersten haben mögen zuthail werden? Welche jetzt leuchtlich ein jeder kan zur hand bringen. Wie vil fürtreffenliche *Opera* in allen *Scientiis*, seind zuvor zu grund gangen, welche mit großen Gelt wider zu erkauffen weren, wans nur müglich were? Selbige hetten auf dise weis wol mögen[f] erhalten werden. Aber alles anders hindan gesetzt, Ich will nur ettlichs von der Sonnen und Mon, als deren die[g] *Astronomia, Optica* und *Physica* sich annemen, anmelden.

---

[a] *Margin*: per … ain. [b] *Deleted*: unrecht ist bewegt. [c] Here the seventh sheet begins. [d] *Added in the margin at a later date*: und zum schirm. [e] *Added in the margin at a later date*: zur belaidigung und schädigung. [f] *Added above the line*: mögen. [g] *Added above the line*: die. *Deleted*: Scientia.

---

207 After several antecedents, the identification of Ophir with Peru was promoted mainly by Benito Arias Montano (1527–1598), the famous editor of the Antwerp Polyglot Bible, in his book *Phaleg*. See Arias Montano (1572, f. A2v). On this and its relation with transoceanic navigation and the important issue of monogenism, see Gliozzi (1977, 147–174), Shalev (2003, 69–71). In 1619, Christoph Besold, the eminent jurist and colleague of Maestlin at the University of Tübingen, published his *De novo orbe coniectanea*, where he opposed the identification of Ophir with Peru and, relying mainly on José de Acosta's *Historia natural y moral de las Indias* (see Acosta (1590), I, 13; Besold quoted from a French translation), argued that the biblical Ophir was located towards eastern India, from where it is very unlikely that a navigation towards the West Indies, that is, present-day Peru, would have been attempted. See Besold (1619), pp. 11–12.—Maestlin again referred to Ophir in the

word from the metathesis of letters.[207] They must have also travelled through the torrid zone aboard these ships according to every map of land and sea, for they entered the Red Sea and still returned home with rich wares unscathed by either leg of their journey. It remains to be said that Aristotle was wrongly swayed by the severe heat in the torrid zone during his own time. This may be all borne in good order, however, for it does his philosophy, and what he otherwise taught well, no disservice in the slightest.

What countless other arts and objects were invented after the age of Aristotle we cannot say. 49 How many more have been invented in the last 200 years (without reckoning any further)? Among them should certainly be counted muskets, useful on the one hand while highly harmful on the other, which may well come to the defense and protection of some but far more frequently are employed offensively, for the unfortunate purposes of harm and injury. How many millions of people have innocently and tragically lost their lives through them since their advent? By contrast, the highly valuable and venerable art of the printing press has been founded. With what great labor and toil (not to mention cost) did we once have to transcribe books before they could have barely reached the hundredth person? Now anyone can promptly pick up their own.[208] How many excellent works in all the sciences have previously perished that would have to be purchased again at great expense, if only it were possible? Yet the same works might well have been preserved in this way. But setting all other things aside, I wish only to address some things about the Sun and Moon, which are the object of astronomy, optics, and natural philosophy.

---

same terms in the text he wrote in 1621 as a rejoinder to the 1616 condemnation by the Holy Office of the motion of the Earth and the prohibition (until corrected) of *De revolutionibus*. Discovered by C. Frisch and published in his edition of Kepler's *Opera omnia* (vol. i, pp. 56–58), Frisch thought that this text—now edited with a French translation in Copernicus (2015), vol. iii, pp. 654–661—was related to the projected new edition of *De revolutionibus;* see note 228 below. The text had to be added to the end of Maestlin's preface to his edition of Rheticus' *Narratio prima* incorporated into the 1596 *Mysterium cosmographicum* if Rheticus' work were also to be included in the new edition. In the end, it was never published. Kepler had responded to the banning of Copernicus' work in 1619 with his *Admonitio ad Bibliopolas exteros, praesertim Italos* (*Warning to foreign booksellers, especially Italians*), collected in *JKGW*, vi, pp. 543 f.

208 Maestlin presents the theme, a commonplace in the period, of the two canonical inventions (firearms and print), arguing for the superiority of moderns over ancients. For a similar passage in Kepler (who is less critical of the 'machinae bellicae'), see Kepler (1606, 330). We thank F. Seck for the reference.

1. Es haben die Alte *Philosophi* an dem Mon gemerckt und gelehrt, das er mit diser unser Erden ein große Verwandtschafft habe, als (under andern) sein Liecht, welchs er so wol als die Erd von der Sonnen hatt, ist nit so gar durch aus lautter und hell, daraus zu vermutten, das sein *corpus* nit so subtil rund, sonder wie die Erd mit Bergen umbgeben sey,[a] etc. Darumb haben sie den Mon genennet *Terram Aetheream*, die himlische Erden, *Macrob. lib.* 1. *cap.* 11 et 19. *Somn. Scip. Plut. lib.* 2, *cap.* 30. *de plac. Philos. et de facie Lunae. Cicero* 4. *Acad. quaest.* und auch *Aristoteles lib. de Animalibus*, wie solches *Averroes* allegiert (dann selbigs Buch nit mehr vorhanden) in *Serm. de Subst. orbis cap.* 2. *lib.* 1. *de caelo*
50 *comment.* 16. *et lib.* 2. *comment.* | 32. *et* 42. Von des Mons und der Erden[b] Astronomischen und Optischen Verwandtschafften *vide Disput. nostram de Planetarum apparentibus irregularitatib. thes.* 68. *usque* 96. *et thes.* 152. Hingegen wirt gelehrt, gleich wie das Element des Waßers ist vil subtiler, und deswegen sein *superficies* zur glatten runde mehr formiert, als die Erde, und gleicher gestalt das Element des Luffts vil subtiler als das Waßer: Also seyen in gleicher proportion die obere *corpora* der Mon, die Sonn und Sternen gegen den undern. Deswegen müßte der Mon, und vil mehr die Sonn, vil subtiler und glatter poliert und rund sein, als kein Spiegel glaß oder Stahel baliert sein kan. Wer gibt alhie den Ausschlag? Antwort: Dieweil auch dis orts der Anfang *à sensu* und vom ὅτι, wie droben gesagt, soll gemacht werden, so geben den Ausschlag die vor wenig Jahren herfür gebrachte *perspicilla, Tubi Optici,* oder Rohrspiegel, wie sie von ettlichen genennet werden. Von denen hatt freylich *Aristoteles* zu seinen Zeiten auch nichts gewüßt. Dise Spiegel geben klar zu sehen, was fern und weit von dannen ist, als were es zu aller nächst, vor dem gesicht. Sie sind zwar langst erfunden worden, dann also schreibt *Cornel. Agrippa* vor 80. oder 90.

---

[a] *Added above the line*: sey. [b] *Added above the line*: des Mons … Erden. *Deleted*: iren.

---

209 Macrobius (2001–2003), I, 11, 7: "Denique illam [Lunam] aetheriam terram physici vocaverunt, et habitatores eius lunares populos nuncuparunt"; 19, 8: "… lunam, quae luce propria caret et de sole mutuatur".

210 Plutarch (1516, f. 17[r]): "Pythagorei glebaceam, propterea videri lunam, quod habitetur quomodo et haec terra nostra". *De facie in orbe Lunae* repeatedly argues for the borrowed light of the Moon and her similarity to the Earth; see also Plutarch (1957, 923A, 926 C, 929 A–B, 931 D and 935 C: "we are far from thinking that the moon, because she is a celestial earth, is a body without soul and mind"); for a recent reappraisal of the subject, see Fabbri (2012).

211 Cicero (1933), *Academica*, II (Lucullus), xxxix, 123, on Xenophanes' opinion, quoted below, p. 51.

212 Averroes (1986, 93): "And since it is apparent in the case of the moon that it is dense and dark by its very nature and that it receives its light from another planet, namely the sun, Aristotle asserts concerning it in the book *De Animalibus* that its nature is generically more

1. The ancient philosophers observed the Moon and taught that it has a great affinity with our Earth, including (among other things) its light, which it draws, just like the Earth, from the Sun, but it is not blatant and bright, from which we may presume that its body is not so perfectly round, but like the Earth covered with mountains, and so forth. That is why they referred to the Moon as "*terra aetherea*," the celestial earth, in Macrobius, Book 1, Chaps. 11 and 19 of *Commentarii in Somnium Scipionis*,[209] Plutarch, Book 2, Chap. 30 of *De placitis philosophorum* and *De facie quae in orbe lunae apparet*,[210] Cicero in Book 2 (Lucullus) of the *Academicae quaestiones*,[211] and even Aristotle in the *Liber de animalibus*, as Averroes alleges (for the same book is no longer extant) in *Sermo de substantia orbis*, Chap. 2,[212] Book 1 of the *De caelo*, commentary 16, and Book 2, commentaries | 32 and 42.[213] On the astronomical and opti- 50
cal affinities of the Earth and Moon, see our *Disputatio de multivariis motuum planetarum in coelo apparentibus irregularitatibus*, Theses 68 to 96 and 152.[214] On the other hand, it is said that just as the element of water is much subtler, and thus its surface more perfectly round in shape, than the Earth, the element of air is much subtler than water by the same measure; the superior bodies, Moon, Sun, and stars, should also stand in the same proportion with the lower ones. That is why the Moon, and still more the Sun, should be much subtler and more smoothly polished and round than any mirror, glass or steel can be polished. Who can reach a resolution to the matter? Answer: since we should also start here with our senses and concerning the question of *quid*, as said above,[215] a resolution may be achieved by the recently introduced spyglasses, optical tubes, or telescopes, as they are called by many.[216] Aristotle certainly did not know anything about them in his own day. These looking glasses show clearly what is far away from them as if it were up close and in front of one's face. They were, in fact, invented a long time ago, for Cornelius Agrippa thus

---

like the nature of earth than like that of the other stars". As noted by A. Hyman, in Arabic and Hebrew philosophy this title includes the *Historia animalium*, the *De partibus animalium*, and the *De generatione animalium*. For the Latin translation probably used by Maestlin, see Aristotle (1560, 323A).

213 Aristotle (1560, 16B, 121D, 132E). For a recent edition, see Averroes (2003, I, 29; II, 324, 355).

214 Maestlin (1606, 40–52, 89–90). All previous references in notes 209–213 are presented in thesis 96 and the commentary on it.

215 See pp. 18, 36 above.

216 As is well known, the *perspicillum* or telescope was introduced into astronomical observation by Galileo, who published his first results in the *Sidereus nuncius* (Venice 1610). On Maestlin's hostility towards Galileo, whom he considers a plagiarist, see Bucciantini, Camerota, Giudice (2012, 120f.). This reference to Agrippa and the telescope is absent from the rough copy; it was introduced for the first time on p. 3 in the intermediary papers.

Jaren, *de Vanitate Scientiarum cap.* 26. *de Specularia.* Ich kan, sagt er,[a] spiegel machen, mit welchen man bey Sonnen schein, alles was sie bescheint, auch an weit entlegnen orten, auf drey oder vier meil wegs, völliglich sehen kan. Man hatts aber in geheim (wie wol gläublich und auch zu beweisen) gehalten. Nun komen sie jetzt gleichsam als gar von newem widerumb herfür. Doch, damit die jenige, welche sie wider aufgebracht, publiciert, und jedermeniglich gegünnet, ires billichen Lobs von Mir nit beraubt werden, so waiß ich wol, das es mehrmalen zween oder mehr auf ainen Huffschlag seind komen, und was der ein erfunden, der ander selbiges auch, ohn empfangne Anlaitung vom ersten, sonder gleich so wol von im selbs, erfunden hatt. Das aber dise newlichere, und, des *Agrippae* Spiegel ainerlay seyen, ist daraus zu erachten, das sie ainerlay laisten.

Wer nun mit disen Spiegeln hinauf in Himmel bis zu dem Mon steiget, das ist, den Mon dardurch mit fleis beschawet, der wirt gewar, und sihet gar deutlich, das im Mon, so wol als hieunden auf Erden, vil und hohe Berg, kleinere doch an der Zahl mehr gegen Mittag, großere gegen Mittnacht, sampt tieffen Thälern seind. Das es aber rechte und warhafftige Gebürg seien, ist augen-
51 scheinlich zu sehen, | dann der Sonnen schein erraicht und bescheinet je und je einen Berg hinder dem andern, das thal aber, so entzwischen ist, wirt übergangen. Item zwischen disen Gebürgen laßen sich, sonderlich im Vollschein, wan der Mon den Sonnenschein völlig empfanget, gleichsam goldfarbe äderlin, sehen, welche sich mit den Waßerflüßen auf Erden vergleichen. Hieraus mag man nicht untaugenlich colligieren und erachten, das die trübere Mackeln im Mon sich vergleichen mit den Wälden, Wildtnußen, und ungehewren Gebürgen auf Erden, welche gleichermaßen schwartz, dunckel oder blaw scheinen: Was aber am Mon liechter und hell ist, vergleiche sich mit den großern Waßern, Seen oder Meer, oder auch mit den schönen Awen, blach und Bawfeldern, von welcher hellem glantz einem, der es von einem hohen Berg, bey Sonnen schein zu Sommers zeit, übersihet, das Gesicht vergehet (Ettliche kehrens umb, und halten, die trübe Mackeln im Mon, vergleichen sich mit den Waßern, und das liechte und helle mit den Bergen oder Wälden etc. Aber Ich kan mir

[a] *Margin*: sagt er.

217 See Agrippa von Nettesheim, *De incertitudine et vanitate scientiarum et artium*, (Chapter 26, *De Specularia*), reedited in *Opera*, vol. II, Lyon, s. d., 60–61: "Et novi ego fabricare specula, in quibus lucente Sole, omnia quaecunque illius radiis illustrantur per remotissima spatia, puta trium aut quatuor milliarium, plenissimè conspiciuntur".

218 Maestlin's animosity towards Galileo, though implicit, is manifest here.

219 This passage on the surface of the Moon (pp. 50–51) is absent from the rough copy, but already appears in the intermediary papers (p. 3).

wrote about them 80 or 90 years ago in *De vanitate scientiarum*, Chap. 26, "On Mirrors." "I can," he says,[217] "make mirrors with which one can see completely everything that lies in the light of the Sun, even in far removed places three or four miles away." However, they were kept secret (as is quite plausible and can also be proven). Now, they are emerging again as if anew. Yet in order that those who again discovered, published, and shared their discovery with everyone not be deprived by me of their due praise, I well know that more often two or more may have taken the very same step, and that what one invented another did as well without receiving instructions from the first but equally on his own. That this more recent looking glass and that of Agrippa are one and the same, however, can be gathered from the fact that they function the same.[218]

Whoever ascends to the heavens with this looking glass all the way up to the Moon, that is, [whoever] diligently examines the Moon through it, will realize and see quite clearly that on the Moon, just as here below on earth, there are many high mountains, smaller in number more toward the South, greater [in number] toward the North, along with deep valleys.[219] Indeed, it is clear to see that they are really and truly mountains, | for every now and then the 51
light of the Sun reaches and illuminates one mountain behind another but passes over the valley that lies between them. In addition, one may see between these mountains, especially in full illumination, when the Moon receives the full light of the Sun, a golden vein, as it were, that can be compared with the channels of water on earth. From this, one may not unreasonably gather and presume that the dark spots on the Moon may be compared with the forests, wilderness, and enormous mountains on the Earth that appear black, gray, or blue in equal measure; while what is bright and luminous on the Moon, on the other hand, may be compared with the great bodies of water, lakes or seas,[220] or the beautiful meadows, open fields, and farming plots, whose brilliant light strikes the face of anyone who overlooks them from a tall mountain beneath the bright summer sun (Some turn the whole thing around and hold that the dark spots on the Moon compare with bodies of water, while the bright and luminous parts compare with forests or mountains,[221] and so forth. But I cannot

---

220 That was the opinion of Giordano Bruno before the invention of the telescope; see Bruno (2006[2], 201–203, 209).

221 This is Galileo's opinion as stated in his *Sidereus nuncius* (1610); see Galilei (1989, 43): "Thus, if anyone wanted to resuscitate the old opinion of the Pythagoreans that the Moon is, as it were, another Earth, its brighter part would represent the land surface while its darker part would more appropriately represent the water surface". Kepler adopted this

nicht einbilden, wie, was trüb ist, solte hell scheinen, und was hell ist, solte schwartz, trüb oder blaw scheinen). Und dieweil nit vermuttlich, das der Allweise Gott habe auch im Mon Berg und Thal, vergebens geschaffen, *Deus et Natura*, sprechen die *Philosophi, nihil faciunt frustrà*, Gott und die Natur thun, wärcken und schaffen[a] nichts umb sonst oder vergebens: So möchte man nit unbillich noch weitter argumentieren, das nämlich daselbsten auch weren *alterationes rerum, et inde etiam generationes et corruptiones*, etc. Wann die alte *philosophi* dis so augenscheinlich, wie jetzt wir, durch solche[b] Optische Spiegel hetten sehen könden, sie weren zweiffels ohn in iren meinungen noch steiffer gesterckt worden, da sie darfür hielten, es seyen im Mon vil Stät und Berg, wie *Xenophanes* lehrt, und werde umb und umb mit vernünfftigen[c] Menschen und unvernünfftigen[d] Thieren, gleich so wol als die Erd bey uns, bewohnet, auch mit vil schönern Gewächss (bäumen und früchten) gezieret, wie des *Pythagorae* Nachfolger sagten, warzu solte sonst jenes on dises? etc.

2. Es bezeugen auch dise Optische Rhorspiegel, das die Sonn nicht so gar *exquisit* und *perfect* circkelrund abgeträhet ist, sonder ir circumferentz ist luckecht, ungleich, und gegen unser Erden zu rechnen, eben so wol als sie, mit
52 Berg und Thälern umbgeben. | Solches hab Ich, und mit mir ettlich andere *Mathematum Studiosi*, mehrmal mit sonderm Fleis observiert. Dann als wir anfänglich uns deßen verwunderten, und vermeineten, das Gesicht, und klarer Sonnen glantz hette uns betrogen, haben wir, allem Argwon einigs Ihrthumbs nottürfftiglich fürzukommen, zu underschidlichen Zeitten, früe, ehe die Sonn zu hell gescheinet, bisweilen in dinnem Nebel, mit abgewechselten ettlichen deren Rohrspiegeln, observiert, und jeder sein observation besonder auf Papeir abgezaichnet, Aber selbige, als wirs zusamen hielten, allmal gleichförmig befunden. In disen *observationibus* werde Ich desto mehr confirmiert, nach dem mir seidher *Ioh. Fabricii Narratio de Maculis in Sole* zu lesen fürkommen, darinnen Er meldet, das sein Vatter David Fabricius, bisweilen (villeicht vor mir) auch Ungleichait und Stuffen in der Sonnen Rand gemercket habe.

[a] *Margin*: wärkhen und schaffen. [b] *Margin*: solche. *Deleted*: dise. [c] *Margin*: vernünfftigen. [d] *Margin*: unvernünfftigen.

concept and abandoned the earlier one in his *Dissertatio cum nuncio sidereo* (1610). See Kepler (1965, 26–28). Maestlin doubtless knew their opinions but preferred not to mention them.

222 Maxim or *auctoritas* n° 18 on Aristotle's *De caelo* (I, 4, 271a 33) according to Hamesse (1974, 161).

223 See Cicero (1933, *Academica*, II, *Lucullus*, XXXIX, 123): "Xenophanes says that the moon is inhabited and is a land of many cities".

imagine how something dark should appear light, while something light should appear black, gray, or blue). And since it is not probable that the All-Wise God would have created in vain mountains and valleys also on the Moon—"God and Nature," the philosophers state, "do nothing in vain,"[222] which is to say that God and Nature act, perform, and produce nothing for no reason or in vain—one might thus further argue with good reason that the same would apply to any form of change [*alterationes rerum*], and thus to generation and corruption as well, and so forth. If the ancient philosophers could have seen this as clearly as we do now through such optical looking glasses, they would have no doubt been fortified in their views even more firmly, since they believed there are many cities and mountains on the Moon, as Xenophanes teaches,[223] and that it is inhabited all around by rational people and irrational animals, in the same way as we witness here on earth, and adorned with much more beautiful plants (trees and fruits), as the followers of Pythagoras said, for why else should this be without the other? And so forth.[224]

2. These optical spyglasses also indicate that the Sun is not so exquisitely and perfectly round in shape, but its circumference is full of holes, uneven, and, measured against our Earth, just as covered by mountains and valleys. | I, and 52
many other mathematicians with me, have observed this several times with particular industry. When we were first struck by this and supposed that our eyes and the light of the bright Sun had fooled us, we proceeded, under suspicion of some error that had inevitably occurred, to observe at different times in the morning, before the Sun began to shine too brightly, at times through the thick fog, with a series of several different spyglasses, each one sketching his own observation separately on paper, only to find every time that we had agreed upon exactly the same thing. I am all the more convinced by these observations, after which I since had the opportunity to read Johannes Fabricius' *Narratio de maculis in sole*, in which he reports that his father, David Fabricius, sometimes (perhaps before me) also noted unevenness and ridges around the edge of the Sun.[225]

---

224 Cf. Plutarch (1516, f. 17^{r}, cit. note 210 above), (1957, 937 D–E): "I should like before that to hear about the beings that are said to dwell on the moon. [...] If it is not possible, the assertion that the moon is an earth is itself absurd, for she would then appear to have come into existence vainly and to no purpose, neither bringing forth fruit nor providing for men of some kind an origin, an abode, and a means of life, the purposes for which this earth of ours came into being".

225 The lines "In disen *observationibus* ... gemercket habe" are written almost at the bottom of the page, after the following paragraph; see note 226 below. The reference is to the first printed report on sunspots, that by Johannes Fabricius (1587–1616), son of the more

Item, welches nit weniger zu verwundern, dise Optische Rhorspiegel geben weiter zu erkhennen, das umb die Sonnen schwartze Mackeln, klein und groß, jetzt mehr jetzt weniger, herumb von Aufgang gegen Nidergang sich volvieren, und verrucken. Sie laßen sich mehrerthails ettlich tag sehen, Anfänglich am rand der Sonnen gegen Aufgang, folgenden tagen kommen sie eintweders gegen dem *centro*, und dann hinüber gegen Nidergang zum andern end, oder aber ziehen[a] neben dem *Centro* auf beeden seitten hin, doch bleiben sie alle auf einer Straß von Aufgang nicht abwegs, sonder schlechtlich dem Nidergang zu.

Weil nun so vil Newes täglich herfür kompt, darvon *Aristoteles* und andere
53 nichts gewüßt, inen auch darvon nie ge|traumet hatt, solte darumb *Aristotelis Philosophia* desto geringer zu schätzen,[b] oder gar hinzuwerffen sein? Mit nichten, sonder man soll sie in billichen Ehren halten, und Gott darfür dancken. Was aber weitter hinzukompt, soll man freylich auch nit ausschlagen, sonder Gottes unaussprechliche Weisheit, Gutte, und Allmacht nur desto mehr preisen. Deswegen hette Keckermann, und andere seins gleichen, der unnotwendigen klag wol sollen überhebt sein, und vil mehr gedencken, das noch zu unsern Zeitten, wie Seneca *lib.* 7 *quaest. Nat.* sagt, vil sachen verborgen seind, welche unsere Nachkomme, wan man unser nicht mehr gedencken wirt, erfinden werden. Die Natur eröffnet ire verborgne Schätz gleich als Hailigtum, *arcana et sacra*, nicht auf ein mal, sonder allgemählich. Sie sparet auf die Nachkomme, und künfftige Zeit auch ettwas. Verwunderstu dich, spricht Er, das die Weißhait ire Werck noch nit vollendet hat? Hatt doch die Boshait sich noch nit gar herfürgelaßen. Man erdenckt ja alle tag ettwas News, so eintweders zu nürgents nutzet, oder aber nur zur üppigkait, stolz, hoffart, schwelgerey, und schaden

[a] *Margin*: ziehen. [b] *Deleted*: sein.

famous David Fabricius (1564–1617); see Fabricius (1611, f. C2[v]): "Visus est [Sol] non negligendâ interdum etiam circa margines inaequalitate et asperitate, quam et parenti meo Davidi Fabricio obscurius quidem animadversam ex ipsius intellexi ad me datis literis". For Fabricius' observations of sunspots, see p. 55 below.

226 After this paragraph Maestlin first wrote the following lines: "In disen *observationibus* werde Ich destomehr confirmiert, in dem ich hernacher vernommen, das auch andere *Mathematici* solchen ungleichen und luckechten Rand an der Sonnen (villeicht vor mir) gemerckt haben, wie dann *Johannes Fabricius* solches von seinem Vatter David *Fabricio* in *Narratione de Mac*". Afterwards, he substituted these lines with the similar ones found above this paragraph following a vertical stroke.

227 For a similar passage in accordance with the celestial location of comets, see pp. 43–44 above.

In addition, [and] what is no less astounding, these optical spyglasses further reveal that around the Sun black spots big and small, fewer and more, revolve from East to West, changing from place to place. Most can be seen for several days, at first on the edge of the Sun toward the East, in the following days either moving toward the center and then over to the other end toward the West or slowing near the middle on both sides, but they all remain on the same route from the East, never diverting from their direct course toward the West.[226]

Since there are now so many new things that appear every day about which Aristotle and others knew nothing or even dreamed |, should Aristotle's phi- 53
losophy therefore be all the less appreciated or thrown away entirely? Not at all, rather one should hold it with due respect and give thanks to God for it.[227] Whatever else is added to it one should certainly not reject, either, but only praise all the more God's inexpressible wisdom, excellence, and omnipotence.[228] That is why Keckermann and others like him should have been kept from airing their unnecessary complaint and rather recalled that even in our age, as Seneca says in the *Naturales quaestiones*, Book 7, many things remain hidden that our descendants shall discover after we are forgotten.[229] Nature discloses her hidden treasure in the same way as a sanctuary her secret and sacred things, not all at once but gradually. She sets something aside for successors and a future age. "Are you surprised," He says, "that wisdom has still not completed her work?" Yet evil has still not run its full course, either. We think up something new every day that is either of no use to anyone or simply serves the pursuit of luxury, pride, arrogance, debauchery, and injury. Why should

228 These considerations by Maestlin on the omnipotence and wisdom of God, well beyond the capacity of the human mind, run parallel to those Maestlin deploys in his 1621 addition to Rheticus' *Narratio prima* in the second edition of the *Mysterium cosmographicum* in reply to Brahe's criticism of the enormous empty space between Saturn and the sphere of the fixed stars required by the annual motion of the Earth and the absence of observable annual parallax in the stars. See Kepler (1621, 114f.). It remains an open question to what extent this expanded critique of Brahe could have been part of Maestlin's intended contribution to the new edition of the *De revolutionibus* that was ultimately not realized. Interestingly, Maestlin's reply to Brahe's arguments also provides a refutation of the argument by Francesco Ingoli in his *De situ et Quiete Terrae contra Copernici Systema disputatio*; see Bucciantini-Camerota (2009, 171–172). Though Ingoli's *Disputatio* was not published, a copy reached Kepler in 1617. Kepler's *Responsio ad Ingoli*, written in the same year, also remained unpublished. The fact that Maestlin does not mention Ingoli (no letters between him and Kepler are extant from this period) seems to indicate that he was unaware of his *Disputatio*, since he would not have otherwise failed to reply to the Roman theologian as well.

229 Seneca (1972, VII, 30, 5): "Multa venientis aevi populus ignota nobis sciet; multa saeculis tunc futuris cum memoria nostri exoleverit reservantur".

dienet. Warumb solte, was mehr zur Ehr Gottes, sein Weishait und Allmacht desto mehr zu erkennen, undergetruckt werden? Hett Keckermann *doctrinam Parallaxeon* beßer gestudiert, so würde er über *Galaxiam*, oder die Jacobstraß, und über die Cometen so hefftig nit zürnen, das sie ire Stell und wohnungen im Himmel haben wöllen. So ist es auch meniglich, und zwar ime Keckermann selbsten, wol bewußt, das *doctrina Meteorologica* noch nicht *omnibus numeris absoluta*, volkommenlich, erörtert ist, deßen sich *Aristoteles* selbs, wie droben angezaigt, beklaget.

230 Seneca (1972, VII, 30, 6–32, 1): "Non semel quaedam sacra traduntur: Eleusin servat quod ostendat revisentibus; rerum natura sacra sua non semel tradit. Initiatos nos credimus, in vestibulo eius haeremus. Illa arcana non promiscue nec omnibus patent; reducta et interiore sacrario clausa sunt, ex quibus aliud haec aetas, aliud quae post nos subibit aspiciet. [...] adhuc in processu vitia sunt. Invenit luxuria aliquid novi, in quod insaniat [...]. Nondum satis robur omne proiecimus [...]. Miraris si nondum sapientia omne opus suum implevit?". These passages from Seneca's *Naturales quaestiones* were frequently used by

another endeavor to glorify God, to discern His wisdom and omnipotence all the better, be oppressed?[230] Had Keckermann studied the doctrine of parallax better, he would not have become so enraged over the galaxy, or St. Jacob's Street, and comets for wishing to establish their dwelling place in the heavens. Thus, it has also been well known to many, including Keckermann himself, that meteorological doctrine is still not completely or entirely discussed, about which Aristotle himself complains, as shown above.[231]

---

the *novatores* to bestow authority on their innovative theses in cometary theory and cosmology. Cf. for exemple Kepler (1619, 132): "Seneca Nat. Quaest. lib. 6, cap. 26: Erit qui demonstret aliquando, in quibus Cometae partibus erret, cur tam seducti a caeteris eant, quanti qualesque sint. Contenti simus inventis: aliquid veritati et posteri conferant".

231 See pp. 36–37, 44 above. The telescope, then, as a new wing, complements the geometrical doctrine of parallax and helps to construct a complete and true visual experience of the heavens that refutes their presumed immutability, adding further support to the celestial location of comets.

54 **CAP. IX**

Ettliche fragen, die Cometen in gemein, und was inen anhengig, betreffend

Auf erzehlte Puncten werden vil und mancherlay *obiectiones*, Gegenwürff und fragen fürgebracht, under denen dise folgende villeicht nit die geringste sind.

I. Und erstlich (damit ich an dem, was zu letst ist auf die Baan komme, anfang) ist die frag, für Ains.

1. Ob die schwartze Mackeln, so in der Sonnen observiert worden, in oder bey der Sonnen selbs, oder ob sie nur *phasmata* und erscheinungen im Lufft seyen? Antwort. Ettliche woltens zwar gern im Lufft aufhalten, sonderlich welche die Cometen so ungern im Himmel leiden mögen, ob sie villeicht auch sonsten alle *alterationes* und Enderungen aus der *aetherea regione* abschaffen köndten. Aber *parallaxeon doctrina* erweiset das widerspil. Dann wie dise Mackeln sich am Morgen früe erzaigen, eben am selben ort, es seye gleich oben, unden, oder mitten in der Sonnen, finden sie sich auch zu Mittag, zu Abendt, und umb der Sonnen undergang, ohn allein was ir *motus proprius* austregt, welches nimmer mehr sein köndte, wan es nur erscheinungen und *meteora* im Lufft weren, dann ire *parallaxes*, welche am Morgen und Abendt ire *loca apparentia à veris*, jedes mal aufs wenigst ain grad, das weren zu beeden orten, zween grad, verrucken, würden sie nit an ainem ort so gewis stehn laßen, seitenmal der Sonnen *Diameter* nur auf[a] ein halben grad sich erstrecket. Darumb ists gewis, das dise Mackeln nit hieunden im Lufft, sonder droben, in, bey, oder umb die Sonnen selbs seind.

2. Darnach wirt auch gefragt, ob dise Mackeln an der Sonnen aller erst *in hac Mundi Senecta*, da die gantze Welt sich zum Undergang naiget, als besondere Wunder und Zornzaichen, und künfftiger ding Vorbotten, gleich wie die
55 Cometen, zu erscheinen haben angefangen? Oder ob sie je und allweg | immerdar an der Sonnen geweßt seyen? Antwort. Von disen Mackeln hatt zuvor und ehe die newe Optische Rhorspiegel aufkommen, niemandts zu sagen gewüßt. Der Erste aber, der sie gemerckt, so vil ich in erfahrung gebracht, ist *Johannes*

---

[a] *Added above the line*: auf.

---

232 The title of Chapter 9 and pages 54–57 appear in a first draft on pp. 4–6 of the intermediary papers (with the title on p. 5).

233 The first question is whether sunspots are celestial or sublunary phenomena and, as a consequence, whether mutability shall be extended to the heavens or affects only the sublunary region. The following question, whether the spots are actually on the Sun itself or revolve around it, is addressed later in point 3 (pp. 56–57).

234 Copernicus estimated the diameter of the Sun to be 33′ 54″ at the minimum distance and 31′ 48″ at the maximum; see Copernicus (1992), IV, 21, 208.

Chapter 9 54
Several Questions Concerning Comets in General, and What Follows From Them[232]

On the above said points, we shall submit many and sundry objections (*objectiones*) and questions, among which the following ones are perhaps not the least.

I. And first (so that I may start with what is last on the preceding list) is the question for no. 1.

1. Whether the black spots thus observed in the Sun are in or on the Sun, or are only apparitions [*phasmata*] and appearances in the air.[233] Answer. Many would prefer, in fact, that they abide in the air, especially those who may suffer from even the thought of comets in the heavens, if they could perhaps also abolish all other forms of alteration and change from the ethereal region. But the doctrine of parallax proves the contrary. For these spots may be found in the very same place at midday, in the evening, and around sunset precisely where they appear early in the morning, no matter whether they are above, below, or in the middle of the Sun, excepting only the extent of their proper motion, which could never be the case if they were appearances and *meteora* in the air, for their parallax, which would each time distinguish between their apparent and actual positions in the morning and evening by at least 1°, or at two positions by 2°, the spots would never remain so precisely at the same place, since the diameter of the Sun spans only $^1/_2$°.[234] It is therefore certain that these spots are not here below in the air, but up above, in, on, or around the Sun itself.

2. Next, we shall also ask whether these spots on the Sun first began to appear at this advanced age of the world, when it inclines toward absolute ruin,[235] as special marvels and signs of wrath, and forerunners of future things, in the same way as comets. Or were they already and at every time | there on the Sun? 55
Answer. Before the advent of those new optical spyglasses, no one knew what to say about these spots. Yet the first one to notice them, as far as I know, is

235 As a faithful Lutheran, Maestlin was firmly convinced that the world was soon end and that the second coming of Christ was at hand; for earlier statements in the same direction, see Granada (2007, 105, 109, 113) with respect to the comet of 1580 and the nova of 1572. The imminence of the end of the world was also an argument in Maestlin's rejection of the calendar reform; on this, see Hamel (2017, 248f.). However, as the following lines (p. 55) clearly show, this conviction did not prevent him from rejecting the view that sunspots were exclusive to this final period of the world's existence. Cf. also p. 56 for their interpretation as natural and ordinary phenomena in the Sun.

Fabricius, herrn *Davidis Fabricii, Phrysii, filius*, dazu mal *Medicinae Candidatus*. Disem, als Er solche Spiegel nicht nur auf Erden gebraucht, sonder auch, gleichsam Lusts wegen, die himmlische Liechter, den Mon und dann auch die Sonnen beschawet, sind solche Mackeln ins gesicht kommen, welche Er zu erst seinem Vatter, hernach auch andern angemeldet, die es mit vilfältigem observieren gleicher maßen befunden, also das nun mehr kein Zweifel darein zu setzen ist.

Das aber dise Mackeln an der Sonnen nichts newes seyen, halt Ich seye daher zu erweisen, dieweil man in alten *historiis* dergleichen zwar nit vil, doch[a] ettwas findet. Zur Zeit *Caroli Magni*, Königs in *orientali et occidentali Francia*, im franckenland und franckreich, des Ersten Teutschen Kaisers, ist in der Sonnen, ein wenig über dem *centro*, ein kleine schwartze Mackel, acht tag lang, ettliche schreiben siben, ettliche neun tag, gesehen worden, wie solches zu lesen ist, sonderlich bey Eginardo (*Caroli Magni Alumno, Epistolarum Magistro*, und seinem[b] Dochtermann) in *vita Caroli Magni*[c] und bey *Annonio Monacho*. Andere welche deßen auch gedencken, habens, wie aus allen umbständen abzunemmen ist, von disen ausgeschriben. Dis *phaenomenon* hatt bey manchen seltzame gedancken erweckt, mehrerthails, Als *Annonius, Author Annalium Francicorum, Sabellicus, Polydorus Virgilius*, und andere, habens entlich für *Mercurii stellam* gehalten. Also ist auch wenig Jar zuvor, ehe die Optische Rhorspiegel sind in gemein bekandt worden, ein solches *phaenomenon* und schwartze Mackel in der Sonnen observiert, gleicher[d] gestalt für *Mercurium* gehalten, und von meniglich[e] geglaubt worden.

Andere, Als *Eginardus*, und *Author Fasciculi Temporum* habens für ein besonders wunderzaichen, und andeittung des bald darauff erfolgenden tödtlich Abgangs *Caroli Magni*, gehalten. Selbigen hab auch Ich, in *Disput. de Planetarum irreg: appar: thesi* 98 zugestimmet, und es nit für *Mercurium*, son-

---

[a] *Added in the margin*: zwar … doch. [b] *Added above the line*: seinem. *Deleted*: hernacher. [c] *Margin*: *in vita Caroli Magni*. [d] *Deleted*: und. [e] *Deleted*: wie noch von mir.

---

236 Maestlin refers to Fabricius (1611). See p. 52 above.

237 Fabricius (1611, $C3^r$–$D1^v$).

238 *Margin*: *in vita Caroli Magni*. See Einhard (ca. 770–840), *Vita Caroli Magni*, xxxii: "in sole macula quaedam atri coloris septem dierum spatio visa". Cf. Einhard (1880, § 32): "Very many omens had portended his approaching end, a fact that he had recognized as well as others. Eclipses both of the sun and moon were very frequent during the last three years of his life, and a black spot was visible on the sun for the space of seven days".

239 Annonius Monachus Benedictinus, *De regum procerumque Francorum origine*, Paris, 1514: "Nam et stella Mercurii, xvi Kal. Aprilis, visa est in sole quasi pro una macula nigra paululum superius medio centro eiusdem syderis: quae a nobis octo dies conspicitur: sed quando primum intravit vel exivit nubibus impedientibus minime notare potuimus".

Johannes Fabricius, son of David Fabricius, of Frisia, at that time a doctoral candidate of medicine.[236] When he trained these looking glasses not only on the Earth, but also, as if out of temptation, on the celestial lights, the Moon, and then also the Sun, he first saw those spots and, after telling his father, told others who through manifold observations confirmed the same thing, so there is no longer any doubt about it.[237]

That these spots on the Sun are nothing new, however, I think may be proven from the fact that one finds in the chronicles of history, though not many, certainly some. In the time of Charlemagne, King of Eastern and Western Francia, in Frankland and France, and the first German Emperor, a small black spot was seen in the Sun, slightly above the center, over the course of eight days—some say seven, others say nine—as can be read especially in Einhard (ward of Charlemagne, Master of Letters, and his son-in-law) in the *Vita Caroli Magni*,[238] as well as the work of Annonius Monachus.[239] Others who also recall it have written about these spots, as may be presumed from all circumstances. This phenomenon has awakened strange thoughts in some, most of whom, such as Annonius, author of the *Annales Francicorum*, Sabellicus, Polydorus Virgilius,[240] and others, ultimately identified it as the planet Mercury. Even just a few years earlier, before the optical spyglasses became common knowledge, such a phenomenon or black spot observed in the Sun was identified the same way as Mercury and been believed by everyone.[241]

Others, such as Einhard[242] and the author of the *Fasciculi temporum*,[243] identified it as a special marvel and sign of the swiftly impending death of Charlemagne. I accepted the same thing in the *Disputatio de multivariis motuum planetarum in coelo apparentibus irregularitatibus*, Thesis 98,[244] and determined that it was not Mercury, but, like that other one, an | extraordinary, 56

240 Marcus Antonius Coccius Sabellicus, *Enneades sive Rhapsodia historiarum, Tertia pars, ab inclinatione romani imperii usque ad annum 1504*, Paris, 1509, fol. lviii^v: "visa est etiam Mercurii stella: decimosexto Calendas Aprilis in medio sole velut macula quaedam nigra"; Polydorus Virgilius, *De prodigiis libri tres*, Basel, 1533, lib. II, p. 110: "Anno salutis DCCLXXXVIII [...] Mercurii stella visa est in medio Sole, velut macula nigra".

241 Maestlin is most probably thinking of Kepler, with whom he had maintained a discussion on this matter. Cf. Introduction, Chapter 8 and note 124.

242 See note 238 above.

243 Wernerius Rolevinck, *Fasciculus temporum, vel Chronica ab initio mundi*, [Paris] 1512, p. lxxv^r: "Signa multa precesserunt mortem gloriosi et sancti imperatoris Karoli. [...] Apparuit per 7. dies macula nigri coloris in sole".

244 Maestlin (1606, 54–58), where these references to historical sources (and several others) are already present. After *thesi 98* is deleted: *beyfall gethon* (*gleichsam ein Coryphaeus*).

56 der wie jene auch[a] für ein | ungewohnlichs *prodigiosum et portentosum meteorum* aufgenommen. Zu welcher mainung mich billiche und erhebliche Ursachen bewegten, Als, das dise Mackel war 7. 8. oder 9. tag under der Sonnen gesehen worden, da doch *Mercurius*, wann er dahin kompt, in ettlich wenig stunden, *directus* in 14. oder 16. *retrogradus* aber in 6. oder 7. stunden hindurch streichet. Nach dem nun durch die Optische Spiegel kund und offenbar ist, das solche Mackeln, bisweilen aine, bisweilen zwo, drey, oder noch mehr, auch ohn den *Mercurium*, sich sehen laßen, so ist hiemit aller argwohn[b] so wol des *Mercurii* halben, als eins newen *Portenti* und wunderscheins, gefallen, sonder daraus ist abzunemmen, das solche Mackeln jeder Zeit und für und für geweßt seindt, seittenmal sie auch vor diser Zeit seind, doch *obscurè*, gemerckt worden. Aber niemandt hatts darfür erkhennet, auch seidther niemand deren geachtet, oder in die Sonnen so scharpf zu sehen jemals begert.

3. Ferner, wirt hie gefragt, Was dise schwartze Masen oder Mackeln in der Sonnen seyen? und, ob sie in der Sonnen selbs, wie die Mackeln im Mon, hafften? oder ob sie von der Sonnen abgesöndert, umb sie herumb, wie die Wolcken
57 ob |[c] der Erden, schweben? Antwort. Dis, halte Ich, werde keins Menschen Vernunfft ergründen, die Sonn solches zu erforschen steht uns vil zu hoch. Ob aber dis und dergleichen dahin zu ziehen und zu rechnen seye, da es haißt: **Die himmel sind nit reyn für Gott**, und, **Sihe, der Mon scheinet noch nit, und die Stern sind noch nicht reyn für seinen Augen;**[d] gib ich mehr verständigern zu bedencken. Doch ist menschlichem fürwütz,[e] so fern selbiger nicht unzeittlich, sonder zum lob der Allmacht und Weißhait Gottes gebraucht wirt, nit verbotten, disen geheimnußen ettwas weitters nach zu gedencken. Als: Dieweil nit nur dise Mackeln zum thail größer, als die größte Stern, erscheinen, sonder

---

[a] *Margin*: wie jene auch. [b] *Deleted:* des *Mercurii* halben gefallen. [c] Here the eighth sheet begins. [d] *Margin*: *Iob* 15. v. 15 et 25. v. 5. [e] Grimm *Wörterbuch*: "wiszbegierde, neugierde, verwunderung, staunenswerthes, wunder".

---

245 Maestlin (1606, 54): "Istud autem phaenomenon non ad Mercurium, nedum inter passiones Planetarum Astronomicis legibus circumseptas, recensendum, sed portentosis Meteoris annumerandum esse, indubitatum apud nos est".

246 See Maestlin (1606, 54–58) for a longer presentation of the following reasons.

247 The following passage extending over 14 lines has been vigorously deleted: *Darnach machet mich irr, das die historici die Zeit solcher erscheinung so ungleich aufgezaichnet haben, Sie melden zwar den 17. tag Martii, 16 Cal. April. Ettliche aber, welches doch der warhait gar nit gemeß, setzens in Jar Christi 778. Andere ins Jar 809. Eginardus und Fasciculus temporum, zaigen an, das in den drey letzten Jaren vor Caroli Magni absterben, seer vil wunderzaichen vorher seyen vergangen, das auch Er, Carolus Magnus, selbs vernommen hab, sie trewen und verkündigen Ihme das Ende seins Lebens, under selbigen wunderzaichen*

portentous, and prodigious *meteorum*.[245] I was moved to this opinion by strong and substantial causes,[246] such as the fact that this spot was seen upon the Sun for 7, 8, or 9 days, while Mercury, whenever it passes before it, sweeps through in some few hours, directly in 14 hours or 16 hours, or in retrograde motion in 6 or 7 hours.[247] According to what is now known and apparent through the optical looking glasses, that such spots, sometimes one, two, three, or even more in number, also appear without Mercury, so all doubt about the matter, from the passage of Mercury to a new portent and spectacle, has disappeared, but rather we may presume from this that such spots have existed at any and every time, since they were also perceived before our own time, though obscurely. No one realized it before, however, nor did anyone pay attention to them or seek to look so keenly at the Sun until now.

3. Further, we shall ask here what these black masses or spots in the Sun are, and whether they are found on the Sun itself, like the spots on the Moon, or whether they are separate from the Sun, floating around it[248] like the clouds | 57
around the Earth. Answer. This, I believe, shall surpass the height of human reason, for the Sun lies too far away from us to fathom it.[249] However, should this and other such matters still be submitted for our evaluation, since it is said, **"The heavens are not pure for God,"** and **"Behold, even the Moon is not bright and the stars are not pure in his eyes"**,[250] I leave it for more insightful minds to consider. On the other hand, as long as human curiosity is not inopportune, but employed to praise the endless power and wisdom of God, it is not forbidden from reflecting on these secrets a little further. For example, since not only some of these spots appear larger than the largest stars, but also the

---

erzehlen sie auch die erscheinung diser atri coloris schwartzen Mackeln in der Sonnen. Nun aber in allen disen erzehlten Jaren, stellet calculus Astronomicus den Mercurium auf gemeldten 17. Martii, kein mal under die Sonnen, man ziehe dann Ihne gleichsam mit gwalt, und bey dem haar herzu, welches aber er so leichtlich nit zugibt. These lines reproduce in German the second reason given in Latin for rejecting the interpretation of the phenomenon as a solar transit of Mercury. See Maestlin (1606, 55–56).

248 These are respectively the positions of Galileo in his three *Lettere sulle macchie solari* (1612) and of the Jesuit Christoph Scheiner (1573–1650) in his *De maculis solaribus* [...] *accuratior disquisitio* and *Tres Epistulae de maculis solaribus*, both printed in 1612. See Galilei–Scheiner (2010).

249 Maestlin's scepticism contrasts with Galileo's opinion (i.e., that the spots are on the Sun or in its atmosphere) and that of Scheiner (i.e., that the spots are external to the Sun and move around it as satellites). On this issue, see Shea (1970) and more recent Galilei–Scheiner (2010).

250 *Margin*: *Iob* 15. v. 15 et 25. v. 5. Underlining in red.

auch die Berg in dem Mon[a] so scheinbar, das sie hie unden auf Erden, in so überaus großer Distantz, gesehen und gemercket werden; so ist zu erachten, wie groß sie an inen selbs seyen. Darnach dieweil dise Mackeln in so mancherlay formen sich erzaigen, jetzt einzechtig, bald zwo, drey, oder mehr, in großer anzal,[b] ettliche klein, dann groß, zum thail rund, zum thail zu hauf gesamlet, und wider zertrennt, und zerstrewet: möcht hieraus zu schätzen sein, das sie nicht in oder an der Sonnen seyen, sonder umb sie her fahren, dann sonsten bliben sie ainerlay, gleich wie die Mackeln im Mon: sonder haben *analogiam* und ein vergleichung mit den Wolcken, ohn, das jene auf ainer straß bleiben, die Wolcken aber fahren, nach Elementarischer Natur, on alle ordnung ob der Erden hin und wider. Dises und dergleichen mag man coniecturieren und schätzen, doch bleibts darbey, was *Seneca* von Cometen sagt: **Es seind vil sachen, welche im verborgen fürübergehen, und kommen menschlichen Augen nit ins gesicht. Gott hatt der Menschen Augen nit alles geoffenbaret.** Item, **können wir doch denjenigen, ohn welchen nichts ist, nit wißen, der doch der größte ist** (das ist, Gott in seinem göttlichen Weesen, können wir mit unser Vernunfft nimmermehr begreiffen, vil weniger mit unsern Augen sehen, dann er wohnet in einem Liecht, da niemandt hinzukommen kan, welchen
58 kein Mensch gesehen hatt,[c] | noch sehen kan) **warumb verwundern wir uns, das wir ettwan nit wißen, wie es mit einem fewrlin** (oder alhie, mit ettlichen schwarzen Maasen) **beschaffen ist?** etc. Jedoch dieweil der Allweise Gott von solchen Sachen uns ein Blick sehen laßt, so mag man wol erachten, es seyen dise Mackeln, und Berg an Sonn und Mon nicht vergebens, sonder neben und mit inen,[d] auch noch mehr anders geschaffen, dann es ist ja nit zu glauben, das Gott sein Allmacht habe allerdings und volkommen in disem aintzigen *corpore*,

[a] *In the ms.*: der Sonne. See note 251 to the translation. [b] *Added in the margin*: in großer anzal. [c] *Margin*: *1. Timoth.* 6. vers. 26. [sic; 16]. [d] *Margin*: nicht vergebens ... mit inen.

251 In the text, erroneously, "in der Sonnen" (in the Sun). Cf. p. 58 below: "dise Mackeln, und Berg an Sonn und Mon".

252 Accordingly, Maestlin seems more inclined towards Scheiner's opinion and to interpret the spots as revolving around the Sun and not forming part of it. Maestlin also concedes motion to the spots and not the Sun itself, as rotating on its axis. Thus, he does not share the opinion of Galileo, Kepler (already in the *Astronomia nova*) or Fabricius. See Fabricius (1611, $\mathrm{D2^v}$), who referred to Bruno and Kepler as supporters of the Sun's rotation, and rejected the identification of the spots with clouds: "Hinc tamen nonnulli [...] suspicabuntur corporis solaris in loco suo conversionem, quam Iordanus Bruno asseruit, et nuper admodum defendit in suis, quos de Martis motibus edidit, commentariis Kepplerus vir profundae eruditionis et admirandi iudicii [...] Quid enim faciamus ex maculis, nescio, si non in ipso Sole collocemus, an nubes?". At the same time, the comparison of the

mountains on the Moon[251] are so apparent that they are seen and noted here below on Earth at such an extremely great distance, it may thus be gathered how great they are in their own right. Next, since these spots appear in so many different forms, now one, then two, three, or more in greater number, some small, then large, some round, some assembled together and then separated and scattered, we may reckon from this that they are not in or on the Sun but travel around it, for otherwise they would remain the same, just like the spots on the Moon; however, they do have an analogy and comparison with the clouds, except that they remain on the same route, while the clouds travel back and forth about the Earth without any sense of order, according to elementary nature.[252] This and other such matters we may conjecture and consider, but what Seneca says about comets remains: **"There are many other things that pass by in secret and do not appear before our eyes. God has not revealed all things to the eyes of men."**[253] In addition, **"we still cannot know the one without whom there is nothing, who is the greatest"** (that is, we can never comprehend God in his divine being through our reason,[254] much less see him with our eyes, for he lives in a light that no one can reach, and no one has seen[255] | or can see), **"so why are we surprised that we do not com-** 58
pletely understand how a small fire (or, in this case, several black spots) **came about?"**[256] And so forth. However, since the All-Wise God allows us to see a glimpse of such things, one may thus well consider that these spots and mountains on the Sun and Moon were not created in vain, but that among and along with them still more things were created, for we should certainly not believe that God unleashed his endless power only and utterly on this single body, the

regular motion of the spots with the irregularity of the clouds shows a certain continuity of the idea of celestial superiority over the sublunary region. Cf. p. 58, where it is said that spots are adjacent (*neben*) to the Sun, whereas mountains are *on* the Moon.

253 Cf. Seneca (1972), VII, 30, 3: "Quam multa praeter hos per secretum eunt numquam humanis oculis orientia! Neque enim omnia deus homini fecit. Quota pars operis tanti nobis committitur? Ipse qui ista tractat, qui condidit, [...] effugit oculos". Underlining in red.

254 With this sentence Maestlin corrects Seneca, who continued the passage just quoted in perfect accordance with Stoic pantheism: "cogitatione visendus est [Deus]. Multa praeterea cognata numini summo et vicinam sortita potentiam obscura sunt aut fortasse, quod magis mireris, oculos nostros et implent et effugiunt, sive illis tanta subtilitas est quantam consequi acies humana non possit, sive in sanctiore secessu maiestas tanta delituit", Seneca (1972), VII, 30, 4.

255 *Margin*: 1. *Timoth.* 6. vers. 26. [sic; 16].

256 The underlined text (in red) is a new quotation from Seneca. Cf. Seneca (1972), VII, 30, 4: "Quid sit hoc sine quo nihil est scire non possumus, et miramur si quos igniculos parum novimus".

in der Erden, da wir wohnen, ausgelehret, sonder was Er alhie an dise Erden (welche gegen dem Hohen Himmel nur wie ein Püncklin ist) gewendet, solt er ime nichts weitters vorbehalten haben, und alle Wunder und thatten (wiewol sie nit zu zehlen sind, *Iob* 5. v. 9. *et cap.* 9. v. 10. und man kaum ein füncklein davon erkennen kan. *Syr.* 42. v. 23.)[a] was er hie geschaffen, nicht auch dorten an Sonn und Mon, schaffen könden, oder noch größers zu schaffen vermögen? Aber was selbiges *in specie* seye, hatt er uns verhalten, es ist gnug, das wir, wie gesagt, einen Blick darvon sehen, bis **das Er nicht allein die Gestalt der Erden ernewert**, sonder **ein Newen Himmel und Erden schaffet**,[b] und ohne Zweifel auch dannen her **seine Schaf wirt herzu führen, damit ain hirt und ain Heerd werde**.[c] Ihme seye Lob, Ehr und Preiß in Ewigkait.

11.[d] Auf dis[e] was bisher fürgebracht worden, ist jetzt dise frag zu betrachten, welche droben gleich im Anfang des 7 *Cap.* angezogen worden. Nämlich, Wie, oder, ob die *Exhalationes, vel fumi*, die Dämpf, Düfft, oder Rauch, das ist, die jenige Materi daraus die Cometen generiert werden, könden von der Erden so hoch in himmel hinauff, und höher als die Sonn ist, sich erheben? Hierauf zu antworten, ist zu bedencken, In dem man jetzt gewis waißt,[f] das die Cometen im Hohen Himmel, in *regione aetherea*, stehen, und erscheinen, das es gar nit gläublich, sonder (natürlich darvon zu reden) unmüglich ist, das die von
59 der Erden auffahrende Düfft oder Dünst, | solten mit solchem unbegreifflichem Hauffen bis dort hin, on mercklichen Abgang des Erdbodens, kommen könden. Dann Erstlich,[g] Bedencke man die unermeßliche größe dises unsers Cometen, welchen, wie droben gesagt,[h] über die tausent mal tausent Meilen lang, und

---

[a]*Margin*: alle Wunder ... *Syr. 42.* [43] *v.23.* [b]*Deleted*: da man der vorige nit mehr gedencken wirt. [c]John 10: 16. *Margin*: *Psal.* 103 [104]. v. 30 *Isa.* 65. v. 17. 2 *Pet.* 3. v. 13. [d]*In the ms*: 2. [e]In the rough copy (fascicle a, p. 25), a parallel passage to what follows continues the text quoted above, p. 43 and note 182 to the translation: "Uf dises folget gleich ein andere frag, wie konden dan dise lufft und dunste aus welchen die Cometen herkommen, von der erden so hoch in himmel uber sich fahren? etc. Antwort: In dem man jetzt gewis ist, das die Cometen nit in lufft sonder in hohen himmel stehen so ists gar nit gläublich, sonder natürlich davon zu reden, unmüglich, das die Cometen aus den von der erden auffriechenden lufften oder dunsten generirt werden, seittenmal die gantze Erdkugel, welche gegen die höhe und größe der obern himmeln für ein klain punctlin geschätzt wirt, zu eins solchen Cometen grösse [p. 26] solte kaum gros genug sein, sonder es ist der warhait vil änlicher die Cometen haben iren ursprung aus aetherischer materi. Was aber selbige seye, und woher sie her rierhe, laß Ich denjenigen sagen, der disen erstern und all andere Cometen geschaffen hatt, nämlich der allerweisen und allmächtigen Gott. Es seind doch ja noch auch vil andere sachen in der natur, welche unser vernunfft verborgen und zu erkundigen vil zu hoch seindt". A single horizontal line seems to close the text here with this humble acknowledgement of ignorance with respect to God's majesty; nevertheless, a series of lines, including considerations concerning the various lengths of the comet's tail, follow. [f]*Added above the line*: waisst. *Deleted*: ist. [g]*Margin*: *1.* [h]*Margin*: pag.

Earth, where we reside, but what He has done here on the Earth (which is only a small point with respect to the upper heaven), should nothing further have remained for him, and could not every deed and wonder (although they are without number, Job 5:9 and 9:10, and one can scarcely discern a small spark of them, Sir. 42:23),[257] which he performed here not also have been performed there on the Sun and Moon, or even greater things as well? Yet what may be the same in kind [*in specie*] He has kept from us, and it suffices that we, as said above, see a glimpse of it, until "**He not only renews the face of the earth**", but "**creates a new heaven and earth**", and no doubt He shall then "**send his sheep unto us and there shall be one shepherd and one fold**".[258] To Him be glory, honor, and praise forever.

11. With regard to what has been put forward thus far, we must now consider the same question that was raised above, right at the beginning of Chapter 7. Namely, how or whether the *exhalationes vel fumi*, the vapors, exhalations, or fumes, that is to say, that matter from which comets are generated, could rise from the earth so high up in the heavens, even higher than the Sun.[259] In answering this question, we must bear in mind what we now know for certain, that comets appear and abide in the upper heaven, in the ethereal region, so that it is not only incredible, but (speaking naturally) absolutely impossible that the exhalations or vapors | arising from the Earth should be able to 59
ascend all the way there in such an incomprehensible heap without any perceivable loss of land. For, first of all, consider the immeasurable magnitude of the present comet, which, as said above,[260] was more than 1,000,000 miles long

257 Syr. 42. [43] v.23.
258 John 10: 16. *Margin*: Psal. 103 [104]. v. 30 Isa. 65. v. 17. 2 Pet. 3. v. 13. Underlining in red.
259 See Chapter 7 above, p. 26.
260 Maestlin has not given the intended page. See p. 26, in Chapter 6, corresponding to p. 26 in the rough copy, under the heading "Longitudo caudae" in the right margin.

über hundert mal tausent Meilen brait war, Wie groß ist dann sein *Corpus* an im selbs geweßt? Gewißlich die gantze Erdkugel, wie groß sie immer ist, solte doch, wans in lautter *exhalationes* und Dämpf resolviert würde, zu ein so großen scheinende *corpore* kaum groß gnug sein.

Ja, sprichstu, dis ist wol müglich. Dann die gantze Erdkugel ist ein überaus groß *corpus*, etc. Antwort. Die Erdkugel oder gantze Erdkrais, ist, wie wir Menschen sie ansehen, und gegen uns rechnen, seer groß. Vil leut meinen, es könde nichts größers gedacht werden. Sie helt rings herumb *uno circulo magno* 5400. teutsche Meil, und ist 1720. Meil dick, Sie begreifft in irem Umbkrais mächtige große Länder, als *Europam, Africam, Asiam, Americam*, deren jedes in vil underschidliche Königreich und Landtschafften abgethailt wirt, über dis seind noch grausame und weitläuffige Meer, und darinnen onzahlbar vil und große Inseln. Ja, die Erd ist so groß, das auch die höchste Gebürg und Berg dargegen für nichts geachtet werden. Die Hailige Schrifft (in dem sie nach unserm Menschlichen Verstand, oder vil mehr Unverstand sich richtet, und irem Gebrauch nach, mit uns redet, wie wir zu reden pflegen, seitenmal wir je nicht nur, was von himlischen Dingen gesagt wirt, nit verstehn, sonder wir verstehn auch kaum, oder nit, was von Irdischen sachen gesagt wirt) die Hailig Schrifft, sprich Ich, stelt die Erden uns für, als ein groß *Corpus*, dann sie nennets **die Welt**. Die Welt zwar ist und haißt aigentlich, der gantze Bau Himmels und Erden. **Gott,** | sagt Paulus,[a] **hatt die Welt gemacht, und alles was darinnen ist, Er ist ein HERR Himmels und der Erden.** Item, **Gottes unsichtbars Wesen muß man mercken an den Wercken, die er thut an der Welt, die Er geschaffen hatt.** Nichts destoweniger wirt auch die Erden vil mal, **die Welt** genennet, gleichsam sie der größte thail der Welt were. Als *Christus* sagt zu seinen Jüngern, **Gehet hin in alle welt, und prediget das *Evangelium* allen Creaturen.**[b] Und, von dem Weib, welche Ihm das köstliche Nardenwaßer auf sein Haupt gegoßen hatt, sagt er, **Wa dis *Evangelium* gepredigt wirt in der gantzen Welt, da wirt man auch das sagen, zu irer gedächtnus, was sie jetzt gethon hatt.**[c] Wiewol an disen und vil andern orten, durch die Welt, werden verstanden, die Menschen auf Erden oder in der

[a]*Margin*: *Act.* 17. vers. 24. [b]*Margin*: *Marc.* 16. v. 15. [c]*Margin*: *Matth.* 26. v. 13. *Marc.* 14. v. 9.

261 Since a German mile amounts to approximately 6.4 km., this affords a figure of 34,560 km., short of the present estimation of 40,075 km. The visionary Paul Nagel (ca. 1580–1624) arrived at a figure of 5,614 German miles in his posthumous work *Raptus astronomicus, Das ist Astronomische gewisse wharhafftige Prophecey und Weissagung aus dem Ersten, Andern, und Dritten Himmel, wie solche darinn befunden werden*, s. l. 1627, Di^{r}–Eiii^{r}, presenting this figure as the “golden measure” (*goldene Meßstab* or *königliches Instrument*) which pro-

and over 100,000 miles wide; how great, then, was its body in its own right? Certainly, the entire globe of the Earth, as big as it always is, if it were resolved entirely into exhalations and vapors, should scarcely be large enough for such a seemingly great body.

"Yes," you say, "that is entirely possible. For the full globe of the earth is an extremely large body," and so forth. Answer. The globe of the earth or entire world is, as we regard it and reckon with respect to ourselves, very great. Many people believe that nothing greater could be conceived. She measures 5,400 German miles[261] around her circumference and is 1,720 miles thick, she comprises across her surface large, powerful lands like Europe, Africa, Asia, and America, each of which is divided into many different kingdoms and territories, beyond which still lie savage and sprawling seas full of countless many and ample islands. Yes, the Earth is so big that even the highest hills and mountains are reduced to nothing by comparison. Holy Scripture (which conforms to our human intelligence, or rather to our ignorance, and, employing it the same way, speaks to us as we tend to speak,[262] since we not only do not understand anything said about celestial matters, but also barely understand anything, if at all, said about earthly affairs), Holy Scripture, I say, presents the Earth to us as a great body, for it calls it **the world.** Indeed, the world is and is properly called the whole edifice of the heavens and Earth. **God,** says | Paul,[263] **made the world** 60
and everything in it, He is the Lord of heaven and earth. In addition, **God's invisible essence must be discerned from the works that He performs in the world, which He created.**[264] Nevertheless, the Earth is also called the world many times, as if it were the greatest part of the world. As Christ says to his disciples, **Go into all the world and preach the gospel to all creation.**[265] And, of the woman who poured precious spikenard on his head, He says, **Wherever this gospel shall be preached throughout the whole world, there shall also be told what she has just done in memory of her.**[266] In these and many other places, however, by [the word] world is understood the people on earth or in

---

vides the key, in accordance with the cabbalistic numbers in the prophet Daniel, to the entire duration of the world and the start of the eschatological kingdom of Christ in 1624.

262 Reference to the hermeneutic principle of divine accommodation, common to Hebrew and Christian exegesis, according to which God and the Scripture "speak the language of Man". See Funkenstein (1986, 213 ff.) and note 271 below. From this moment, the text has no precedent in the previous drafts.

263 *Margin*: *Act.* 17. vers. 24.

264 Romans, 1:20.

265 *Margin*: *Marc.* 16. v. 15.

266 *Margin*: *Matth.* 26. v. 13. *Marc.* 14. v. 9.

Welt. Als. **Also hatt Gott die Welt geliebet**, etc. und bald darauff: **Gott hatt seinen Son nit gesandt in die Welt, das er die Welt richte, sonder das die Welt durch in selig werde.**[a] So wirt doch vil mal die Erde, und was darauff ist,[b] verstanden. Als, In der Versuchung zaiget der Taüfel auf einem seer hohen Berg, dem Herrn *Christo* **alle Reich der gantzen Welt, in einem Augenblick, und ire Herrligkait.** Item, **Was hülffs den Menschen, so er die gantze Welt gewünne, und verlüre sich selbs**, oder, **neme schaden an seiner Seel.**[c] Es macht auch die H. Schriftt nit so ein überaus großen Underschid zwischen der Erden und dem Himmel. *Isaias* sagt,[d] **Wer mißet die Waßer mit der Faust, und faßet den Himmel mit der Spannen? und begreifft die Erden mit eim Dreyling** (mit dreyen fingern) **und wiget die Berg mit einem Gewicht, und die Hügel mit einer Wage?** Das ist, Alle Meer und Waßer seind Gott ein Faust oder Hand voll: den gantzen himmel kan er überspannen; die gantze Erden mit dreyen fingern aufheben; Er hatts bey der wag ausgewegen, wievil centner jeder Berg halten soll. Aus welchen worten scheinet, als solten der Himmel, die Erd, das Meer, und die Berg, ein solche proportz gegen ainander haben, gleich als da man ain Ding überspannen, das ander mit dreyen fingern fassen, das dritte in der hand halten, das vierdte mit einer handwag abwegen kan. Solche sachen weren je an
61 irer grösse nit so gar wert ab einan|der. Gleicher gestalt wirt in der H. Schrifft der Himmel verglichen und genennet Gottes Stul, und die Erde der Schemel seiner Füß, oder sein Fußbank.[e]

---

[a]*Margin*: *Joh.* 3. v. 16. 17. 1 *Joh.* 4. v. 9. [b]*Added at a later date*: was darauff ist. [c]*Margin*: *Matth.* 4. v. 8. *Luc.* 4. v. 5. *Matth.* 16. v. 16 [26]. *Mar.* 8. v. 36. *Luc.* 9. v. 25. [d]*Margin*: *Isa.* 40. v. 12. [e]*Margin*: *Isa.* 66. v. 1. *Matth* 5. v. 34–35. *Act.* 7. v. 49.

---

267 *Margin*: *Joh.* 3. v. 16. 17. 1 *Joh.* 4. v. 9. Note that Maestlin distinguishes between the cosmological sense of 'world' (the cosmos created by God or "the whole edifice of the heavens and Earth", as mentioned a few lines above) and the usual sense in the Bible and in human colloquial language, where 'world' means, as has just been said, "the people on Earth" ("die Menschen auf Erden"). This seems to be a way of reminding the detractors of the motion of the Earth (not least the Holy Office and the Congregation of the Index) that the former sense cannot be confronted with the categories with which we deal with distances, velocities and quantities of bodies in the human 'world'. Thus, these considerations would be closely related to Maestlin's program of confronting the condemnation of Copernicus' work and the motion of the Earth through the failed edition of *De revolutionibus* and his additions to the second edition of Kepler's *Mysterium cosmographicum*. See notes 228 and 269.

268 *Margin*: *Matth.* 4. v. 8. *Luc.* 4. v. 5. *Matth.* 16. v. 16 [26]. *Mar.* 8. v. 36. *Luc.* 9. v. 25.

269 *Margin*: *Isa.* 40. v. 12. This verse was quoted again by Maestlin as part of his extensive and very important addition to the second edition of the *Mysterium cosmographicum* (see

the world. For example, when it says, **For God so loved the world**, and so forth, and soon thereafter, **God did not send his Son into the world to condemn the world, but that the world might be saved through him.**[267] Yet that is how the Earth and everything on it is often understood. For example, during the temptation the devil shows the Lord Christ from the top of a very tall mountain **all the kingdoms of the entire world at once and their glory**. In addition, **What does it profit a man, if he should gain the whole world and lose himself**, or **bring harm to his own soul.**[268] Holy Scripture does not make such an extremely great difference between the Earth and the heavens. Isaiah says,[269] **Who measures the waters with his fist and marks the heavens by the width of his hand? And who comprehends the Earth by a third** (with three fingers) **and weighs the mountains with a scale and the hills by a balance?** That is to say, all of the seas and waters are a fist- or a handful for God; He can span the entire sky, lift the entire Earth with three fingers; He has weighed on the scales how many centners each mountain should hold. From these words, it seems as though the heavens, Earth, sea, and mountains should possess such a proportion with one another, as if you could span one thing, seize the second with three fingers, hold the third in your hand, and weigh the fourth in the balance of your palm.
Such things would never bear the same value among each other | with respect 61
to their size. For the same reason, in Holy Scripture the heavens are called and compared with the throne of God, while the Earth is known as the resting place for his feet, or his footstool.[270]

Kepler (1621, 113–119)), namely in the edition of Rheticus' *Narratio prima*, p. 114; see also the French translation in Rheticus (1982, 165–167). It is difficult to decide whether this addition was intended by Maestlin to be part of the new edition of the *De revolutionibus* that in 1613 had been projected by Sebastianus Henripetri to replace the 1566 edition. Henripetri approached Kepler to that effect, and the latter communicated the project to Maestlin, who took a keen interest in it. Maestlin wrote to Henripetri proposing not only a simple correction of the errata present in the previous edition, but also the correction of abundant errors of calculation that he had noted in his copy of the *De revolutionibus*, as well as the inclusion of the *Mysterium cosmographicum* with the *Narratio prima*. The loss of Kepler's letters to Maestlin from April 1607 to May 1616 and those of Maestlin himself between September 1616 and March 1620 deprives us of key information on the development of the project, which was finally abandoned, perhaps because of Henripetri's disinterest (the condemnation by Rome and the publication of the *De revolutionibus* by Nicolaus Mulerius in Amsterdam in 1617 could represent a commercial obstacle) and of Kepler himself, who with the publication from 1618 of his *Epitome astronomiae copernicanae* could consider the work of Copernicus already obsolete. On this question, see now Copernicus (2015, vol. i, 612–618).

270 *Margin*: *Isa*. 66. v. 1. *Matth* 5. v. 34–35. *Act*. 7. v. 49.

Aber in disem allem ist der H. Schrifft mainung nit, das sie die aigentliche größe oder proports Himmels und Erden uns fürhalte, sonder sie richtet sich, wie hieoben angedeuttet, nach unserm und eins jeden geringverständigen Menschen Verstand, sonderlich in denjenigen sachen, daran unser ewige Seligkait nit gelegen ist, wie es dann kein Artickel des Glaubens ist, und zur Seelen seligkeit nit notwendig,[a] das ein jeder so püncklich und gerade müßte wißen, wie groß diser Natürliche Himmel, und dargegen die Erde seye, sonder selbiges wirt in andern *Scientiis* und Künsten (welche auch Gottes gaben seind) ergründet und eröffnet.

Wan wir aber die Erden nit nach unserm Augen maß, sonder nach ausweisung gewisen Astronomischen und Geometrischen hierzu aigentlich gehörenden Grundts, und nach der höhe und größe der öbern Sphaeren im Himmel abschätzen, da ist sie, wie offt gemeldet, nur wie ein Püncklin, oder Dipflin, zu rechnen. Dann wie die *Astronomi* aus unfählbarem Geometrischem und Arithmetischem grund beweisen, ist der Sonnen diameter (die uns kaum einer Spannen, oder eins Schuchs brait bedunckt) sechsthalb mal so groß, als die gantze Erde dick ist. Darumb wie die Erdkugel dick ist, 1720. Meil, also ist der Sonnen diameter mehr als neun tausent Meil, und die Sonn selbs ist 166. mal größer als die gantze Erdkugel.

Aus disem ist jetzt abzunemmen, wie vil tausent und aber tausentmal größer, als die Erde, diser unser Comet muß geweßt sein? Wie wolte es dann müglich sein, das aus und von der Erden köndte so gar vil Dampf oder Rauchwerck, zu einen solchen auf ettlich Monat weerenden Cometen gnugsam ausfahren, ohne wol empfündtlichen deren Abgang? Es müßte ja nicht nur ein Berg oder zween, nicht nur gantze Länder verzeeret werden, sonder so auch der gantze Erdbod würde verstieben und verriechen, möchte es doch kaum gnug sein, oder wans schon dis were, würde doch von der Erden nit vil mehr übrig sein.
62 Nun aber bleibt der gantze Erdkraiß mit Waßer und | Landen in vorigem Stand und wesen, und wirt in aller welt kein münderung gespüret. Darumb ists zu schließen, das die Cometen ire Uhrsprüng, substantz und Materi mit nichten aus und von den Düfften, so von Erden auffsteigen, bekommen, sonder anderswoher haben.

[a] *Deleted*: ist.

271 This is the principle of divine accommodation (see note 262 above), by which the Copernican authors (from Rheticus to Bruno, Kepler and Galileo) sought to defend heliocentrism and the motion of the Earth, as well as freedom of research, from the attacks of theologians and Aristotelians.

In all of this, however, it is not the purpose of Holy Scripture to present us with the actual size or proportion of the heavens and Earth, but it conforms, as indicated above, to the modest measure of human intelligence in each and every one of us, especially in those matters on which our eternal salvation does not depend, as there is thus no article of faith, nor is it necessary for the salvation of the soul, that everyone should know so precisely just how big these natural heavens and the Earth, by contrast, are, for these things shall be researched and revealed in other arts and sciences (which are gifts of God as well).[271]

Yet when we measure the Earth not with our eyes, but according to the demonstration of certain astronomical and geometrical principles that properly pertain to it, and estimate according to the distance and magnitude of the upper spheres in the heavens, there it may be reckoned, as often reported, as merely a small point or tiny speck.[272] For as the astronomers prove from an infallible geometrical and arithmetical principle, the diameter of the Sun (which is considered barely a width of the hand or one foot wide) is $6\,^{1}/_{2}$ times larger than the thickness of the entire Earth. Therefore, since the globe of the Earth measures 1,720 miles across, the diameter of the Sun is more than 9,000 miles, and the body of the Sun is 166 times greater than the total globe of the Earth.[273]

We may now deduce from this how many thousands upon thousands of times greater the present comet must have been than the Earth. How would it have then been possible for so very many fumes or vapors to rise in sufficient magnitude from and out of the Earth to form such a comet lasting for several months without any perceivable loss [of matter]? In fact, it would take not only a mountain or two, not only entire lands to be consumed, but the full bulk of the earth would be dissolved and devoured, if that were even enough, and even if it were, not much of the Earth would still be left over. At present, however, the entire globe preserves the same state of being by land and sea, and no degree of diminution is sensed anywhere in the world. It can therefore | be concluded 62
that comets do not derive their origin, substance, or matter in the slightest by or from the exhalations that arise from the Earth, but acquire the above things from elsewhere.

272 Ptolemy (1984), I, 6: "the earth has the ratio of a point to the heavens"; Copernicus (1992), I, 5, 12: "the earth [...] being like a point in relation to the immense heavens".

273 Cf. p. 54 above and note 234.

11. Darnach, die *Exhalationes, fumos*, Düfft, Dunst, oder wie solche materi, daraus die Cometen ire ursprung und weesen haben solten, immer mögen namen haben, selbsten betreffendt, gesetzt, doch mit nichten begeben, das sie mit so unermeßlichem hauffen aus der Erden, on deren verletzung, münderung und Abgang, köndten ausriechen und sich versamlen: So ist jetzt die frag: Ob sie jemals so hoch hinauf kommen seyen? Hierauf ist ex *Opticis simpliciter* mit **Nein** zu antworten. Ursach dieser Antwort ist, dann sie würden auch vor irer entzündung, oder ehe sie zu leuchten anfiengen, droben gesehen werden. Solches aber ist niemals geschehen, oder das es jemals geschehen were erhört worden.

Dises zu verstehen, ist zu bedencken, das der Lufft hie unden bey der Erden, der Sonnen schein nicht nur durch lauffen laßt, sonder ine, wie ein Christall oder glaß, auch faßet, und selbs scheinend wirt. Dann die Elementen sind nit gar von einander abgesöndert, sonder mit und durcheinander vermischet, sonsten würde aus der Erden nichts herfür wachsen: des Waßers würde weder Menschen noch Vich, noch einige Creatur zu genießen haben: In dem Lufft würde nichts leben könden etc. Darumb, damit alles was aus und in der Erden aufgehen, wachsen und sein Krafft haben könde, so ist (vermög des worts des Herren, *Genes.* 1. vers. 11)[a] die Erd mit allerlay warmen, lüfftigen und feuchten Dämpfen, das ist,[b] nicht nur mit waßer und lufft, sonder auch mit fewr vermengt, dann ja auch in den eißkalten Steinen fewr verborgen ist, und selbige die beste fewrstein sind. Also sind die waßer nicht pur und gantz lautter, sonder ettlichs ist sies, anders ist sawr, anders Bitter, schwebelich, gesaltzen,
63 lieblich | unlieblich, hitzig, siedig, warm, kalt, lawe, etc. Gleicher gestalt ists mit dem Element des Luffts bey uns beschaffen, welcher wie er vermenget seye, zeugen die nebel, wolcken, schnee, regen, ungewitter, blitz, donner, hagel, sträl, sturmwind, etc. welches alles herkommet aus vermischung der Dämpfen, so von der Erden und dem Waßer übersich in den Lufft rauchet und steiget. Es werden aber durch solche Düfft und Dämpf nicht nur trübe wolcken und wetter verursacht, welche den himel bedecken, der Sonnen schein aufhalten, und ine nit durchlaßen, sonder den Lufft auch bey schönem Wetter, und hellen himmel ausfüllen, daher dann kommen taw, reiffen, wärme und kälte auf dem boden, Item in der höhe werden daraus wunderseltzame erscheinungen, als Sternschießen, Sternbutzen, fewrzaichen, fewrflammen (deren ettlichs, wie cap. 1. gesagt, bis auf die Erden herunder, bisweilen mit großem schaden, fallen)

---

[a] "Und Gott sprach: Es lasse die Erde aufgehen Gras und Kraut, das sich besame, und fruchtbare Bäume, da ein jeglicher nach seiner Art Frucht trage und habe seinen eigenen Samen bei sich selbst auf Erden. Und es geschah also", Luther's translation. [b] *Deleted*: mit wasser, lufft und fewr.

11. Further, regarding the exhalations, fumes, smoke, vapors, or however such material sources from which comets should draw their nature and origin might be called, given, though not granted in the least, that they could arise from the Earth and assemble in such an immeasurable mass without any damage, diminution, or loss to it, the question now is whether they ever reached that high up. Here, we must answer on the basis of optics simply with **No**. The reason for this answer is that they would have been seen up there even before they were ignited or had begun to shine brightly. This has never happened, however, or we would have heard of such a thing.

To understand this, we must consider that the air here below on earth not only allows the light of the Sun to pass through it, but also takes hold of it like crystal or glass and begins to shine on its own. For the elements are not separate from one another in the slightest but mixed thoroughly together, otherwise nothing would grow out of the Earth: neither man nor livestock, nor any living creature, would have water to enjoy, nothing could live in the air, and so forth. Therefore, in order that everything that [comes] from and in the earth could emerge, grow, and gain its strength, the earth (by virtue of the word of the Lord, Gen. 1:11)[274] was mingled with all sorts of warm, airy, and moist exhalations, that is, not only with air and water, but also with fire, for there is even fire that lies hidden in ice-cold stones, which are, in fact, the best flint stones. Thus, the waters are never entirely pure or sound, but some are sweet, some are
sour, and some are bitter, sulfurous, salty, pleasant |, unpleasant, hot, boiling, 63
warm, cold, tepid, and so forth. The same is true for the element of air around us, which, in what way it may be mingled, is witnessed by fog, clouds, snow, rain, storms, lightning, thunder, hail, squalls, wind storms, and other things, which all come from the mixture of exhalations that exude from the earth and water and rise up into the air. Yet such fumes and vapors not only cause dark clouds and dismal weather that cover the sky, smother the Sun, and prevent it from shining, but they also fill the air during clement weather and clear skies, thus producing dew, frost, cold and warmth in the ground, while on high they bear witness to wondrous appearances such as falling stars, shooting stars, fiery signs, fiery flames (some of which, as said above in Chap. 1,[275] fall all the way down to earth, causing great damage in some cases), *chasmata* [chasms],

274 "And God said: 'Let the earth sprout vegetation, plants yielding seed, and fruit trees bearing fruit in which is their seed, each according to its kind, on the earth'. And it was so".

275 See p. 2 above (in the rough copy).

Chasmata, erschrockliche hellene, da offtermal der Himmel oder ein großer thail deßelben als brennendt, mit grausamem fechten, wütten, durcheinander schießen und fahren, gesehen, ettwan auch schrecklichs krachen[a] gehört wird. Und ob schon solche und dergleichen *meteora* nicht allmal erfolgen, so ist doch der Lufft mit den Dämpfen und Düfften stäts gefüllet, wie dann solches bey klarem Himmel augenscheinlich gemerckt kan werden. Als da die Sonn durch ein fenster oder loch in ein Gemach scheinet, sihet man klar und deuttlich, wie selbiger schein auf dem boden oder an der Wand durch einander schimmert und zwitzert, und zwar solches Zwitzern[b] spüret man auch in dem Lufft selbs. Dises Zwitzern und wuslen ist nichts anders als die hin und wider schwebende Düfft und Dämpf. Sonderlich aber wie der Lufft hieunden dicker seye als in der höhe kan man täglich früe, an der Morgenröte, und spat, an der Abendtröte gleichsam handgreiffig vernemmen. Nämlich, die Sonn, welche, wie droben angezaigt,[c] vil mal größer ist als die Erdkugel, scheinet durch die gantze Welt, das ist, ir schein geht durch den gantzen himel, und auch durch den Lufft, allein[d] ausgenommen, was in der Erden schatten begriffen wirt. Diser Schatten aber ist und haißt die Nacht, und ist[e] das geringste thail der welt, gleich wie die gantze
64 Erd gegen dem himmel nur wie ein pünck|lin zu rechnen ist. Daraus folget, das auch bey stickfinsterer nacht, da die Sonn under dem Horizont ist, ir[f] schein nichts desto weniger durch den Himmel und Lufft auch hie ob dem Horizont, ohn, wie gesagt, was in der Erden Schatten gefaßt wirt, durchgehe. Das aber selbiger Sonnen schein bey nacht nit gesehen wirt, ist dis die Ursach, dieweil der Himmel und gantze *aetherea Regio*, wie auch der obere Lufft, so zart, rein und subtil seind, das sie den Sonnen schein zwar durchlaßen, aber nit faßen noch erleuchtet werden. Dis ist zu mal ein gewiss anzaigung, das bis daselbst hin kein dicke materi, es seyen *vapores, exhalationes, fumi*, Düfft, Dämpf, oder[g] dergleichen, so den Sonnen schein auffangen und erleuchtet werden, kommen kan. Hergegen aber wann die Sonn gegen dem Aufgang sich zu dem horizont nahet, als dann fahet der himmel an weis, hell oder rot werden, und hiemit gehet die Morgenröte auf, und bricht der Tag an. Solches ist ein anzaigung, das danzumal der Sonnen schein über dem Horizont außerhalb der Erden schatten dasjenige thail des Luffts antreffe, welches der Erden vil näher, und deswegen nit so subtil, pur und sauber, wie der obere Lufft ist, sonder mit exhalationen und Dämpfen vermenget, welche den Sonnen schein faßen, halten, und hiemit selbs scheinen könden. Als dann wirt der Lufft allgemählich liecht, bis es heller tag wirt, und zwar ein gutte Zeit zu vor, ehe die Sonn aufgehet.

[a] *Margin*: gesehen … krachen. [b] *Margin*: Zwitzern. [c] *Margin*: wie droben angezaigt. [d] *Added above the line*: allein. *Deleted*: und aber. [e] *Deleted*: diser Schatt. [f] *Deleted*: der Sonnen. [g] *Deleted*: wie sie möchte Nomen haben.

terrifyingly bright, when often the sky or a large portion of it is seen like burning timber, with fearsome darts of fire dashing all over, and even a horrible crackling heard at times. And although such *meteora* and others of the same sort do not occur all the time, the air is still always full of fumes and vapors, as can be plainly seen when the sky is clear. For when the Sun shines through a window or a hole in a room, one can see clearly and plainly how, on the floor or a wall, the light shimmers and shudders all over, and the same quivering, in fact, can even be felt in the air itself. This quaking and quivering is none other than the fumes and vapors floating back and forth. Yet one can examine palpably, as it were, just how much thicker the air is here below than on high each day in the morning at dawn and in the evening at dusk. That is to say, the Sun, which, as shown above,[276] is many times greater than the globe of the Earth, illuminates the entire world, that is, its light passes through the whole heavens, and also through the air, except for what is grasped in the shadow of the Earth. Yet this shadow is known as the night, and it is the smallest part of the world, just as the entire Earth amounts to only a small point | with respect to the heavens. 64
From this, it follows that even in the darkest of night, when the Sun lies below the horizon, its light nevertheless passes through the heavens and air even here above the horizon except, as said above, for what is grasped in the shadow of the Earth. That the same light of the Sun is not seen at night is due to the delicate, pure, and subtle nature of the heavens and whole ethereal region, as well as the upper air, which allow the light of the Sun to pass through without ever being apprehended or illuminated by it. This is, above all, a certain indication that no thick materials can reach all the way up there, whether they are *exhalationes, fumi, vapores*, fumes, vapors, or other things of this sort that are intercepted and illuminated by the light of the Sun. However, when the Sun approaches the horizon near sunrise, the sky then starts to turn white, bright, or red, and then dawn breaks and a new day begins. Such is an indication that at that time the light of the Sun above the horizon, beyond the shadow of the Earth, encounters that part of the air that is far closer to the surface, and therefore not as subtle, pure, or pristine as the upper air, but mingled with exhalations and vapors that could stop and seize the sunshine, and thereby shine on their own. For at that time the air grows gradually lighter until it becomes bright day, and even a good while before the sun ever comes up.

---

276 Cf. p. 61 above.

Durch dises mittel rechnen die *Astronomi* sampt den *Opticis*,[a] und meßen, wie hoch die Dämpf,[b] auch die zärtiste, und subtiliste exhalationen von der Erden in den lufft hinauf kommen konden. Alhazen *de crepusculis, Num.* 6. und *Vitellio, lib.* 10. *prop.* 60. rechnen 51. *miliaria* 47′. 34″. 31‴. oder gar nahendt 52000. *passus*.[c] Das sind 52. (arabische, ettwas geringer, als welsche) meil, deren irer Rechnung nach ein *circulus magnus in terra* begreifft 24000. meilen, und ein grad, 66. 40′ meilen. Aber nach unserer Rechnung, die wir 19. meilen für ainen grad zehlen,[d] erraichet die höchste distantz diser Dämpf noch nit gar 12. teutsche meilen.

65 [e] Solche Rechnung wirt auf dises fundament gegründet. Es haben die *Astronomi* durch vilfältige *observationes* erlernet, das die größte fix stern, so *primae magnitudinis* sind, nach dem sie jährlich ein Zeitlang under der Sonnen verborgen gwest, sich am morgen nit ehe wider sehen laßen, es seye dann die Sonn, ehe es recht heller tag wirt, noch 12. grad under dem horizonten. Also die Stern, so ein wenig kleiner, und *secundae magnitudinis*, komen nit ehe herfür, es seye dann die Sonn noch 13. grad under dem horizont. Gleicher gestalt fordern die Stern, so der dritte größe sind, von der sonnen 14. grad. Eben also werden von grad zu grad[f] denen, so der vierten größe, 15. grad; denen so der fünfften größe, 16. grad; und denen so der sechsten größe, 17. grad zugerechnet. Wann aber die Sonn noch 18. grad (alles *in circulo verticali* gerechnet) under dem Horizonte ist, als dann fanget der Lufft erstmals an sich zu entfärben, doch also, das man nichts desto weniger die aller kleinste Sternlin noch sehen kann. Und von disem termin fahen gemeinlich[g] die *Astronomi* und *Optici* das *Crepusculum*, und[h] den ersten Anfang der nächstfolgenden tags helle, oder die Morgen röten zu rechnen. Eben dise gestalt hatts auch zu abendt nach der Sonnen undergang, wann sie 18. grad hinunder kompt, so ist alle tags helle auch hinunder und verloschen.

Wiewol[i] nun diser termin weit gnug ausgespannet ist (dann die tägliche *observationes* und aufmercken bezeugen gnugsam, das morgens und abendts[j] von dem tagliecht aufs wenigst nichts kan vermerckt werden, wan schon die Son ettwas nähers bey dem Horizont, als 18. grad, ist) jedoch zum überfluß,

[a] *First version*: und die Optici. [b] *Deleted:* vapores. [c] *Deleted*: welches thut nach unsern teutschen meilen zu rechnen 11 2/3 elf [?] zwey drittel teutsche meilen. [d] *Deleted*: rechnen. [e] Here the ninth sheet begins. Page 65v is blank. [f] *Margin*: von grad zu grad. [g] *Margin*: gemeinlich. [h] *Margin*: das Crepusculum, und. [i] *Deleted*: aber. [j] *Deleted*: nicht.

Astronomers and opticians reckon by the above means and measure how high the vapors, including the finest and most subtle exhalations from the Earth could ascend in the air. Alhazen in *De crepusculis*, No. 6, and Witelo [in *Optica*], Book 10, Prop. 60, reckon 51 miles and 47°34′31″, or altogether approaching 52,000 paces.[277] That means 52 (Arab, somewhat shorter than Roman) miles, according to which calculation a great circle around the Earth encompasses 24,000 miles and one degree 6640 miles. Yet according to our calculation, by which we count 19 miles for one degree, the farthest distance of these vapors does not even reach 12 German miles.

The foregoing figure shall be founded on the following basis. Astronomers 65
have learned through manifold observations that each year the largest fixed stars of the first magnitude, after they have been hidden beneath the Sun for a while, do not appear again in the morning unless the Sun, before the day is quite bright, is still 12° below the horizon. Slightly smaller stars of the second magnitude, however, do not emerge again unless the Sun is 13° below the horizon. By the same measure, stars of the third magnitude require 14° from the Sun. And so it goes, degree by degree; thus, for those of the fourth magnitude, the measure is 15°; for those of the fifth magnitude, 16°; and for those of the sixth magnitude, 17°. However, when the Sun is still 18° below the horizon (calculating everything on a vertical circle) and the air first begins to grow faint, one can still see all of the smallest stars in the sky. This is the point astronomers and opticians generally mark to measure dusk and the first onset of the following brightness of day, or dawn. The same thing can be said for the evening after sunset, when the Sun reaches 18° below and the last light of day has also descended and disappeared.

Although this point is now far enough removed (for daily observations and the attentive eye sufficiently indicate that the slightest hint of daylight cannot be discerned in the morning or the evening when the Sun is already somewhat closer to the horizon than 18°), Alhazen sets the target one degree further down

277 See Risnerus (1572) pp. 284, 287 s. (for Alhazen), pp. 452 s. (for Witelo): "Summum ergo, ad quod elevantur vapores secundum ipsorum consistentiam, est minus quam 52000 passuum, ut patere potest perquirenti" (p. 453).

und[a] allem Zweifel fürzukommen, steckt Alhazen das Zil noch umb ein grad weitter hinunder,[b] und setzt 19. grad. Welchem auch *Vitellio*, und aus angedeutter Ursach ettliche *Astronomi* auch folgen.

---

[a] *Deleted*: damit. [b] *Margin*: Alhazen *de crepusc: Num. 1.*

---

278 *Margin*: Alhazen *de crepusc: Num. 1*; cf. Risnerus (1572, 284 and also Chapter 6, pp. 277 f.). *De crepusculis*, attributed to Alhazen, was first printed in 1542, accompanying the work by Petrus Nonius Salaciensis (Pedro Nunes or Nuñez) *De crepusculis liber unus*; see Nonius (1542, part 2, prop.1): "Allacen vero huiusmodi distantiam solis ab horizonte, gradus habere subiecit 19, nulla ratione suffultus. Deinde Vitelo eo iunior, et à quo universum fere ingens illus opus suum de ratione videndi mutuatus est, gradus etiam 19, continere scribit, idque

and submits 19° to be safe and dispel any doubt.[278] This is what Witelo[279] and several astronomers also follow for the reason indicated above.

instrumento armillarum aut tabulis per observationem astronomicam deprehendisse. Denique recentiores omnes praedictam distantiam gradus 18, habere subiiciunt". For the correction of the erroneous attribution to Alhazen, see Sabra (1967, 77), who demonstrates that "it is in fact the work of the Andalusian mathematician Abu ʿAbd Allāh Muḥammad ibn Muʿādh, who lived in the second half of the eleventh century".

279 Cf. Witelo, *Optica* (10, 60), in Risnerus (1572), pp. 452 f., corresponding to the reference given by Nonius in the previous note.

Appendix 1
Can Comets Be Predicted?[a]

[Fascicle 8a, p. 9] An Cometae prognosci possent?

Hie entsteht so wol von allen Cometen, als in sonderhait von disem Cometen mancherlei fragen, deren die eine frag ist, ob man *ex arte et scientia Astrologica* könde die erscheinungen eins ongewohnlichen *meteori*, fürnemlich eins Cometen prognosticieren und zuvor, ehe sie kommen, verkündigen? Antwort. Ich bin zwar kein *Astrologus*, hab mich auch darauff, gewißer ursachen wegen, nie begeben wöllen. Jedoch befinde Ich, das die jenige, welche *Astrologiam prudenter* und mit rechtem verstand und maß tradieren, vil *arcana*, und gleichsam geheimnussen der Natur, fürnemlich was *in regione Etheri* sich zu begeben pflegt oder mag,[b] zu offenbaren wißen, sonderlich wenn sie *intra terminos artis et scientiae* (welches laider nit von allen geschiht) bleiben und nit zu wait schreitten. Aber die *praesagia* und verkündigungen der Cometen betreffend,[c] ist mein einfeltig Bedencken: das was *Astronomica* sind, und Ihren gewißen lauf in der Natur fein haben, kan mal wol wißen, dann man waißt solches gewiß

---

[a] The rough copy presents on p. 9, after line 8, in the middle of the second chapter, a horizontal line from left to right separating the preceding text from what follows. This new text presents on the left the heading "An Cometae prognosci possent?" (see Figure 3 in the Introduction, p. 50) without indicating that a new chapter begins. There is no doubt, however, that this is a new issue, independent from the one discussed in Chapter 2 concerning the appearance of the third comet. This new issue extends over pp. 9–10, since on p. 11 a new section begins under the heading "Motus via", which coincides with Chapter 3 in the clean copy (fascicle 8c); see note 29 to the translation. This discussion about the possibility of predicting comets was probably absent from the lost first pages of the clean copy, dealing with Chapters 1–2 of the treatise. The comparison of the two writings does not offer enough space for this discussion in Chapter 2 of the clean copy. Accordingly, we are inclined to believe that Maestlin intended to deal with the issue in the clean copy in one of the final, unwritten chapters, most probably in one devoted specifically to this significant subject, as well as to the matter—also absent from the unfinished clean copy—of the significance ("Bedeuttung") of the comet, as the first thirteen lines of p. 27 in the rough copy indicate. In all probability, this section on the predictability of comets was the negative verdict Maestlin planned to deliver after Johann Faulhaber, a mathematician in Ulm, pretended to have prophesied the apparition (for 1 September 1618 o.s.) of the first comet of 1618 in a calendar published in 1617. As indicated in the Introduction, Chapter 9, Faulhaber tried to contact Maestlin through his friend Matthäus Beger, a mathematician in Reutlingen who reached out to the Tübingen astronomer. In his letter to Beger of January 1619, Maestlin declared his scepticism over the predictability of comets, as well as his intention to write about the matter in his cometary treatise under preparation. There is thus no doubt that this section in the rough copy of Maestlin's treatise on the comet of 1618 is at least a draft of what he prepared in response to Faulhaber and his pretension of predicting comets. [b] *Margin*: fürnemlich ... mag. [c] *Margin*: betreffend.

Appendix 1

[Fascicle 8a, p. 9] Can comets be predicted?

Here arise various questions about all comets, as well as the present comet in particular, one of which is whether one could prognosticate from the astrological art and science the appearances of an extraordinary *meteorum*, especially a comet, and proclaim them before they appear. Answer. I am certainly not an astrologer, nor have I ever wished to resort to astrology for certain reasons.[280] On the other hand, I find that those who practice astrology prudently and with the proper measure and sense know how to uncover many mysteries and secrets of nature, especially what may or tends to occur in the ethereal region, particularly if they remain within the limits of art and science (which unfortunately is not the case with everyone) and do not stray too far. Yet when it comes to the predictions and proclamations of comets, it is my simple scruple that one can well know those things that are astronomical, as well as the certain course of comets in nature, for we know how to calculate with certainty such things as

280 For similar refusals in his earlier cometary treatises, see Maestlin (1578, 55), (1581, 35).

aus zu rechnen, als *in coelo*[a] das ein planet bisweilen größer scheint, dann zur andern Zeit, welches in Marte vilmal geschehen ist, welcher wegen seiner ongewohnlichen größe, nämlich wan er in *Epicycli et Eccentrici sui perigaeis* sich befindet,[b] mehrmalen ist von gemeinen leutten für ein newen stern gehalten und, neben andern, auch Ich bin darumb ersucht worden. Also Venus so offt sie über 30. grad von der ⊙ kommet, kan bey tag vil Stunden lang gar wol[c] gesehen werden und was dergleichen *Astronomica* seind. Dergleichen auf Erden und in dem Wasser und Lufft geben die *Aspectus stellarum certas irradiationes*, von welchen die undere Elementarische Natur sich reguliert und richtet. Daraus ein verständiger und behutsamer *Astrologus* die veränderungen des Luffts und vil anders prognosticiren kan. Was aber *in coelestibus* und mit astronomischer rechnung *ex motibus* nit mag erraicht werden, zweifelt mir seer, ob ettwas gewißes oder vermutlichs mög astrologisch judiciert werden.

Als v. g. die *Aspectus* ♄ et ☽, oder anderer planeten könden Ihre würkhungen hie unden in der Erden haben. Das aber sie hinauf in den hohen himmel gleicher gestalt nit wirken, kan hieraus abgenommen werden.

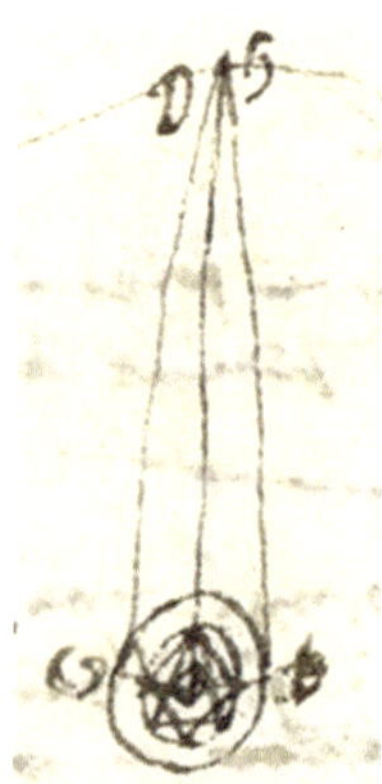

Hiebey gesetzter circulus sol verstanden werden des Mons *sphaera* wie sie die Erden, so mitten ime stehet, umbgibt. ♄ steht in seiner *sphaera*, welche von der letsten, wie *Copernicus* beweißt, schier zweihundert mal, aber nach der gemeinen *Astronomorum* meinung mehr als 200 mal höher stehet als der Mon,

[a] *Margin*: in coelo. [b] *Margin*: nämlich wan … sich befindet. [c] *Margin*: vil Stunden lang gar wol.

281 And, as a consequence, they do not cause or provoke a comet, which may be predicted from the aspects.

when a planet appears larger at times in the heavens than at others, which has often happened in the case of Mars, when it grows significantly in size, that is, when it reaches perigee on its epicycle and eccentric, and is generally held by the common people to be a new star and even I am called upon, among others, for this reason. Thus, every time Venus moves more than 30° in front of the Sun it can be seen quite well for many hours during the day, among other astronomical matters of this sort. In the same way, on earth and in the water and air the aspects of the stars give certain radiations, by which lower elementary nature is directed and ordered. From this, a shrewd and careful astrologer can predict changes in the air and much more. Yet I strongly doubt whether what may not be achieved in celestial matters according to astronomical calculation from the motions may be judged astrologically with any degree of certainty or probability.

For example, the aspects of Saturn and the Moon, or of other planets, could exercise their effects here below on the Earth. That they do not have the same effect on high in the upper heaven,[281] however, can be deduced from the following.

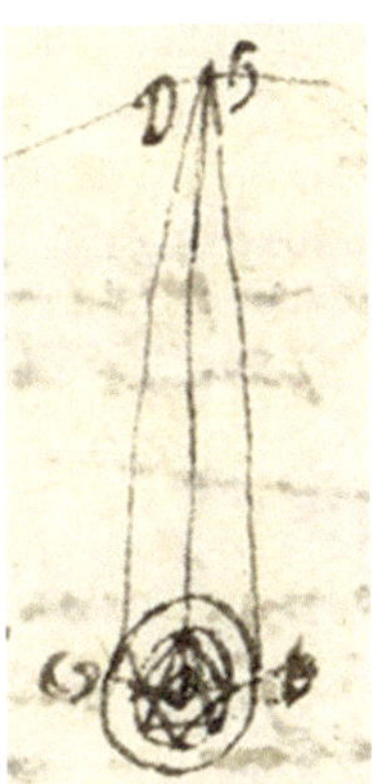

The accompanying circle should be understood as the sphere of the Moon as it surrounds the Earth, which is situated in the middle of it. Saturn is set on its sphere, which, as Copernicus proves, is almost 200 times farther away from the Earth than the Moon,[282] but according to the general consensus of astronomers

282 See *De revolutionibus*, IV, 17 for the distance of the Moon (65 e.r. maximum distance) and V, 9 for the distance of Saturn. See also Van Helden (1985), pp. 46 f., where the maximum distance between the Earth and Saturn amounts to 12,252 e.r.

deswegen wan ♄ und ☽ *in quadrato aspectu* gegen einander stehen, das ist, wan in circulo ☽, oder in circulo altiori seind 90. *gradus*, oder ad *terram* winkelrecht seind, so gibt der *angulus* ad B nur 3. grad. So kan zwar die *influentia* dises *aspectus* under sich in die Erden wircken, aber mit nichten über sich, seitenmal der gantze lauff des Mons durch seinen gantzen *circulum*, gegen den B nur 6 grad austregt. Gleicher gestalt stehts mit den andern planeten auch.[a] Nun schweben die Cometen, wie droben bewisen,[b] nicht *in elementari*, sonder *in aetherea regione*, im hohen himmel, wie solches drunden wider *Artis sententiam* soll erwisen werden. Wie wolte dann einer *ex arte Astrologica* künden aus den Aspecten welche herunder *ad terram*, und mit nichten *ad planetas alios in ipsorum sphaeras*,[c] wircken.

Und ob wol obiiciert mag werden, das ettliche *Astrologi* je und je Cometen[d] prognosticiert und gewissen haben: disen kan zur antwort entgegen gehalten werden, das nach dem von *Anno* 1572. bis 1607, seind 7. Cometen und 2. newe Stern seind erschinen, also haben vil prognosticanten in iren *practicis* auf newe Stern und Cometen andeuttung gethon, deren zwar ettliche es errathen, und dise haben hernach mit Ihren *prognosticis* nit wenig gepranget. Hergegen aber hatts der mehrer thail gefählt, und dise haben hernach von Iren prognosticiren nie vil wert gmacht, sonder sie kondten stillschweigend wol fürüber gehn. Daraus wol mag geschlossen werden, das auch die andere Ihre Cometen nur gleichsam ongefahr[e] errathen haben. Dan es ja geweis,[f] das was übernattürliche sachen seind, deren erkentnußen ist des menschen vernunfft nicht vähig, darumb sie selbige auch nit zu vor zu verkündigen wußten.

[a] *Margin*: gleicher ... auch. [b] *Added above the line at a later moment*: wie droben bewisen. [c] This substitutes a preceding *in suis sphaeris* [later *suas sphaeras*]. Our reading of this sentence is uncertain. [d] *Added above the line*: Cometen. *Deleted zweimal*: mit iren. [e] *Margin*: gleichsam ongefahr. [f] Grimm *Wörterbuch*, vol. 6, col. 5456: "*vereinzelte nebenform zu* gewisz".

283 The demonstration is absent from the preceding pages. As a consequence, the most plausible explanation is that Maestlin intended to present these considerations in one final

is more than 200 times, which is why, when Saturn and the Moon are situated in quadrature with one another, that is, when in the circle of the Moon, or in an upper circle, measures 90° or forms a right angle with the Earth, the angle with B yields only 3°. Thus, the influence of this aspect can act below it on earth but not in the least above it, since the entire course of the Moon over its full circle measures just 6° with respect to B. The same can be said for the other planets as well. Now, comets, as proven above,[283] do not float in the elementary, but in the ethereal region, in the upper heaven, as should be shown below, contrary to common opinion. So how could anyone from the art of astrology pretend to know anything from the aspects, which act here below on earth but not in the least on other planets in their own spheres?

And although it may be objected that several astrologers have at one time predicted comets correctly, it can be held on the contrary in response that, from the years 1572 to 1607, 7 comets and 2 new stars appeared while many prognosticators gave notice in their annual predictions [*practicis*] of new stars and comets, some of which hit the mark and later bragged more than a bit about their prognostications. On the other hand, the greater part missed the mark and then never made much of their prognostication, but could pass entirely over it without another word. It may be concluded from this that even the others only anticipated their comets approximately, as it were. For it is certain that human reason is not capable of comprehending supernatural things, so they could not have known to announce them in advance, either.

---

chapter of the treatise, after the demonstration of the celestial location of comets. This addition contrasts with the following statement in the first writing "wie solches drunden wieder *Artis sententiam* [the common opinion] soll erwisen werden", which implies that Maestlin initially intended to address this question of the predictability of comets in the first chapters, when addressing the question of the location of comets. This explains the presence of this section on predictability in the preliminary pages of the draft.

Appendix 2
Draft of a Letter to Duke Johann Friedrich to Apologize for the Delay in Presenting the Requested Report

[Fascicle 8a, p. 27; lines 14–51. See Figure 6, p. 198]

g⟨nädiger⟩ fue⟨rst⟩ und Herr.

E. f. g. seind all meine eifrigen gebett zu gott für dern und das Haus wirtemberg glückliche wolfart ⟨und⟩ mein underthenige gehorsame Dienst in schuldiger underthenigkeit zuvor.

gn⟨ädiger⟩ f⟨ürst⟩ und Herr

Auf E. f. g. gnädigen erstlich zugetheilten auch etlichmal gnedig widerholten Bevelch[a] (welchen ich jedes mal in gebürender †reuerenz† underthenig empfangen) hab Ich in underthenigkeit dem newlich erschinen und noch alberait am himmel stehenden Cometen,[b] mit sonderm fleiss so nahe mit observiren so lang mit calculiren, und die *observationes* examiniren mich auffs eusserst beflissen.

Das aber nit nur wider e. f. g. gnedig begeren sonder auch ich wider meinen willen ettwas lengers verziehen muoste, bitt E. f. g. ich in aller underthenigkeit Sie wölle solches in ungnaden mich nit entgelten lassen sonder dessen zuvor ursach gnädig vernemmen.[c]

Ich hab aber auch dahin zu bearbeiten befleisen wöllen, das E. f. ⟨g⟩ ich nit nur ein *rude* oder *generale aliquid scriptum* offerirte, sonder welches ettwas was notwendig were so vil müglich[d] begreiffe. Deshalben hab Ich ja vermeint ich wölle dis *Cometae motum non tantum quoad apparentias, à posteriori*, sonder auch *à priori* könte beschreiben und dieweil er ja *aethereus* ist, verhofft *inter sphaerarum coelestium orbes* einen gewissen *orbem*, wie *in anno* 1577 ich selbiges *Cometae sphaeram et orbem in sphaera* gefunden auch sein *motum ex orbe unà cum vera distantia à terra* computiert.[e] Aber *perplexitas motus huius cometae*[f] hatt mich †gelossen†[g] sonderlich dieweil er erstlich ein richtigen *circulum* in seim *motu* observirt, aber letzlich darvon abgewichen, und er ist zwar nit *retrogradus* auch nit gar *stationarius* worden, aber so langsam fortgeschritten

[a] *Deleted*: Mandat. [b] *Margin*: dem newlich … Cometen. [c] *Underlined*: anhören. *Above the line*: vernemmen. [d] *Above the line*: so vil müglich. [e] *Deleted*: abgemessen. [f] *Deleted*: (welcher anfanglich ein richtigen motum ghabt). [g] *Deleted*: bewegt.

Appendix 2
Draft of a Letter to Duke Johann Friedrich to Apologize for the Delay in Presenting the Requested Report

[Fascicle 8a, p. 27; lines 14–51. See Figure 6, p. 198]

Gracious Prince and Lord.

Your princely grace, may all my fervent prayers to God be for the fortune and welfare of you and the house of Württemberg, and may my loyal and obedient service be in humble obedience to you.

Gracious Prince and Lord

In response to the gracious order once given by your princely grace and graciously repeated many times (which I dutifully received each time as a due request), I have, with respect to the comet that recently appeared and still remains in the heavens, made every effort to observe closely, with special diligence, and at length with calculation, and to examine the observations.

That I was obliged to delay it some more time, not only against the gracious wishes of your princely grace but also against my own will, I beg your princely grace with utmost obedience not to allow me to suffer for in disgrace, but to graciously consider the above cause for it.

I have further endeavored to edit the text, however, so that I might offer your princely grace not only something written rudely or generally, but that I may grasp that which is necessary as much as possible. In fact, that is why I supposed that I could rather account for the motion of this comet not only in terms of appearances, *a posteriori*, but also *a priori*, and since it is certainly ethereal, I sought out a certain orb among the orbs of the celestial spheres, just as I had found the sphere and orb of the comet of 1577 and calculated its motion from the orb, together with its actual distance from the Earth.

However, the intricacy of this comet's motion threw me off, especially since it first observed a proper circle in its motion but ultimately deviated from it and became neither retrograde nor entirely stationary, but progressed so slowly that

das er *stationario similis* ist worden.[a] Hiedurch bin Ich in *calculo* verhindert und abgehalten worden, und hinden nach selbigen gar muste fallen lassen. Und ob jetz zwar,[b] da er abgangen, auch von dem vollen monschein verdeckt war,[c] also das Ich gewartet, er were nun mehr gar verschwunden, er sich aber noch imerdar sehen laßt (wie Ich dem gestrigen tag und[d] am morgen früe und zu abendt spatt ine[e] gesehen hab) hette Ich lieber sein *motum* nit *punctatim* sonder[f] *ulterius ab initio ad finem* wollen beschreiben, ob doch hiezwischen sein *orbis* wer zu †ergrunden† geweßt. Bitt E. f. g. nochmals gantz underthänig, sie wölle nit ungnaden nach uns ansehen, sonder das *scriptum* mit gnaden annemen, auch mein gn⟨ädig⟩er fürst und herr gnädig sein und bleiben.[g]

Walthe[h] E. f. g. dem gnädigen Gott in seinen vatterlichen schirm, dan uber mich zu gnaden underthenig bevelhend.

9. Febr. 1619

---

[a] *Deleted*: gar gleich. Cf. Chapter III, p. 11: "bis er zuletzt schier gar still stehen wollen, und *stationarius* zu werden das Ansehen bekommen". [b] *Deleted*: dieweil Ich dieser Comet. [c] *Deleted*: ist. [d] *Deleted*: früe. [e] *Above the line*: ine. [f] *Margin*: nit … sonder. [g] *Deleted*: noch. *Margin*: auch … bleiben. [h] "Sei vorhanden". Cf. Grimm, *Wörterbuch*, vol. 27, col. 1391.

---

284 Cf. Chapter 3, p. 11: "until it finally appeared to stand almost entirely still and become stationary".

it soon grew similar to stationary.[284] As a result, I was hindered and halted in my calculation and then had to let the whole thing drop entirely. And even though it has now diminished and been covered by the full light of the Moon,[285] so that I expected it would have disappeared even more completely, it can still be seen in the sky (as far as I last saw it yesterday in the early morning and late evening),[286] yet I would have preferred rather to trace its motion not concisely, but more from beginning to end, even though its orb would then have to be examined further. I humbly beg your princely grace once again not to look upon me with disfavor, but to graciously receive my text, and to graciously remain my benevolent prince and lord.

May your princely grace enjoy the grace of God in his paternal protection, then humbly beseeching his grace upon me.

9 February 1619

285 There was a full Moon from 19 January (old style). See Kepler (1616, 103). We thank Friedrich Seck for this information.

286 After the full moon, the comet could be seen again from 25 January. According to this, the date of 9 February at the bottom of the letter was probably written after Maestlin had finished the letter. We may recall that Maestlin claims to have seen the comet in February and even March.

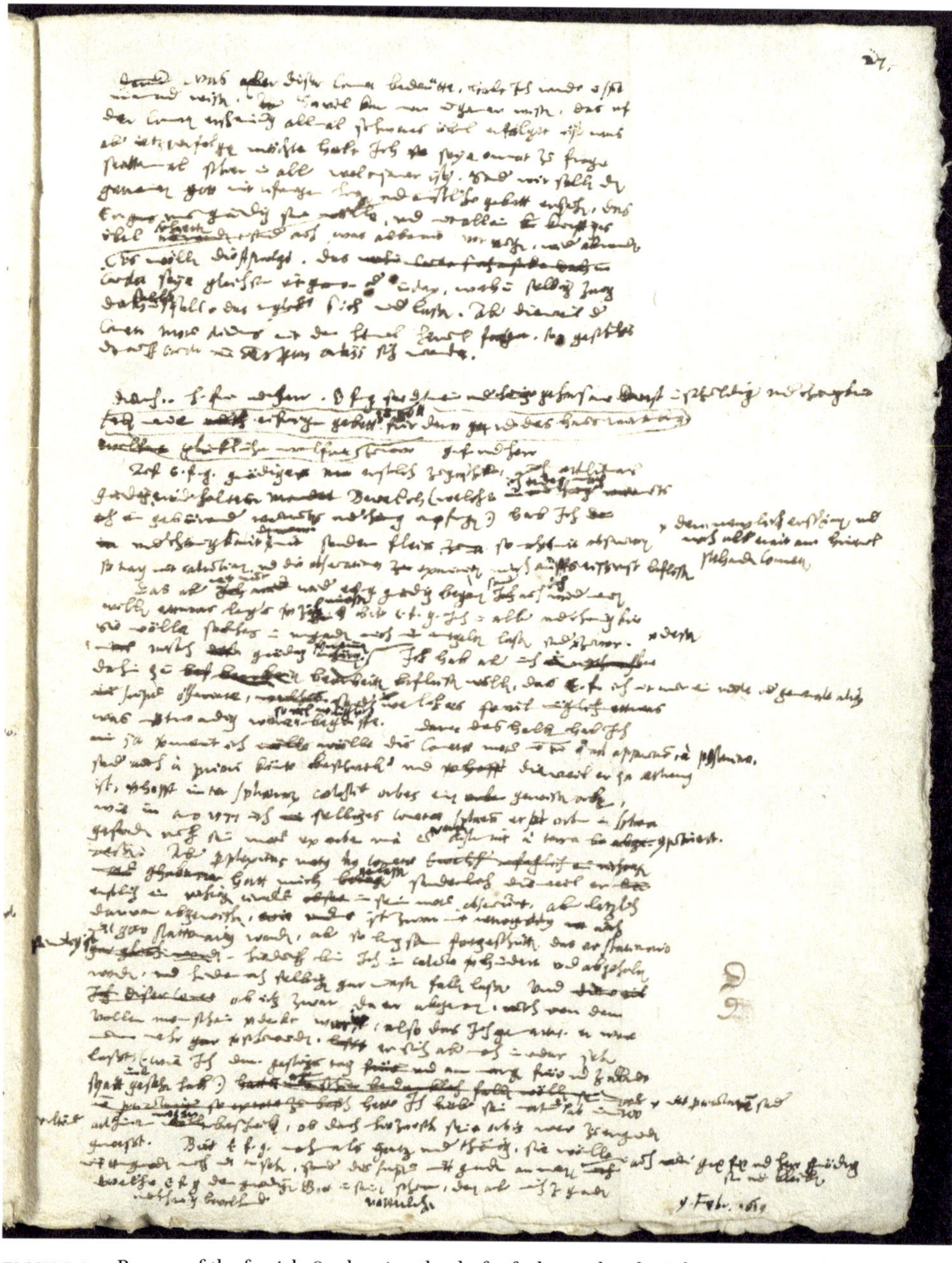

FIGURE 6 Page 27 of the fascicle 8a showing the draft of a letter, dated 9 February 1619, to the Duke

Bibliography

Acosta, J. de (1590). *Historia natural y moral de las Indias*. Seville: Iuan de León.

Agrippa von Nettesheim, H.C. (1527). *De incertitudine et vanitate scientiarum et artium* [Cologne], reprint in *Opera*, vol. II, Lyon s. d.: Per Beringos Fratres.

al-Biṭrūjī (1971). *On the Principles of Astronomy*, ed. by B.R. Goldstein, 2 vols. New Haven—London: Yale University Press.

Amerbach, V. (1545). *Ioannis Iovani Pontani Liber de meteoris (cum interpretatione Vitti Amerbachii)*. Strasbourg: Crato Mylius.

Apianus, P. (1540). *Astronomicum Caesareum*. Ingolstadt: In aedibus nostris [Petrus Apianus].

Arias Montano, B. (1572). *Phaleg, sive, De gentium sedibus primis, orbisque terrae situ liber*. Antwerp: Christophorus Plantinus.

Aristotle (1560). *Omnia quae extant opera*, vol. V: *De coelo et mundo libri IIII. De generatione et corruptione libri II. Averrois De substantia orbis*. Venice: Comin da Trino.

Aristotle (1984). *Meteorology*, translated by E.W. Webster, in *The Complete Works*. The Revised Oxford Translation edited by J. Barnes. Princeton, NJ: Princeton University Press.

Aristotle (1984b). *Posterior Analytics*, translated by J. Barnes, in *The Complete Works*. The Revised Oxford Translation edited by J. Barnes. Princeton, NJ: Princeton University Press.

Aristotle (1984c), *On the Soul*, translated by J.A. Smith, in *The Complete Works*. The Revised Oxford Translation edited by J. Barnes. Princeton, NJ: Princeton University Press.

Averroes (1986). *De substantia orbis*, Critical Edition of the Hebrew Text with English Translation and Commentary, by Arthur Hyman. Cambrige MA–Jerusalem: The Medieval Academy of America and The Israel Academy of Sciences and Humanities.

Averroes (2003). *Commentum magnum super libro De celo et mundo Aristotelis*, edited by F.J. Carmody and R. Arnzen, 2 vols. Leuven: Peters.

Baldini, U. (1992). *Legem impone subactis: Studi su filosofía e scienza dei gesuiti in Italia, 1540–1632*. Rome: Bulzoni.

Barker, P. (1993). The Optical Theory of Comets from Apian to Kepler. *Physis* 30, 1–25.

Barker, P.—Goldstein, B.R. (1998). Realism and Instrumentalism in Sixteenth-Century Astronomy: A Reappraisal. *Perspectives on Science* 6, 232–258.

Barnes, R.B. (1988). *Prophecy and Gnosis: Apocalypticism in the Wake of the Lutheran Reformation*. Stanford, CA: University of Stanford Press.

Barnes, R.B. (2016). *Astrology and Reformation*. Oxford: Oxford University Press.

Beger, M. (1619). *Problema Astronomicum: Die Situs der Sternen Planetarum oder Cometarum zu observirn ohne Instrumenta, allein mit einem geraden Lineal oder Faden.*

Welcher Modus fast gebraucht wirdt von dem fürtrefflichen Mathematico und Astronomo M. Michaeli Maestlino [...] Dem Ehrnvesten, Wolgeachten und Fürnehmen Herrn Johann Faulhabern, bestellten Mathematico etc. zu Ulm, Seinem sondern guten Herrn und Freund, printed with Faulhaber (1618–1619).

Besold, C. (1619). *De novo orbe coniectanea*. Tübingen: Apud Joh. Alexandrum Cellium.

Besomi, O. and Camerota, M. (2000). *Galileo e il Parnaso tychonico: Un capitolo inedito del dibattito sulle comete tra finzione letteraria e trattazione scientifica*. Florence: Olschki.

Betsch, G. (1996). Südwestdeutsche Mathematici aus dem Kreis um Michael Mästlin. In Irmgard Hantsche (Ed.), *Der "Mathematicus": Zur Entwicklung und Bedeutung einer neuen Berufsgruppe in der Zeit Gerhard Mercators*, Duisburger Mercator Studien, Volume 4, pp. 121–150. Bochum: Universitätsverlag Dr. N. Brockmeyer.

Betsch, G. (2002). Parerga Maestlini. In Gerhard Betsch and Jürgen Hamel (Eds.), *Zwischen Copernicus und Kepler—M. Michael Maestlinus Mathematicus Goeppingensis 1550–1631*, (Acta Historica Astronomiae 17), pp. 141–156. Frankfurt: Harri Deutsch.

Bianchi, L.—Randi, E. (1990). *Le verità dissonanti. Aristotele alla fine del Medioevo*. Rome–Bari: Laterza.

Bocarro Francês, M. (2009). *Tratado dos cometas que apareceram em Novembro passado de 1618*. Facsimile edition by H. Leitão. Lisbon: Biblioteca Nacional de Portugal.

Boner, P.J. (2013). *Kepler's Cosmological Synthesis: Astrology, Mechanism and the Soul*. Leiden: Brill.

Brahe, T. (1588). *De mundi aetherei recentioribus phaenomenis*, Uraniburg. In Tycho Brahe, *Opera omnia*, ed. by J.L.E. Dreyer, vol. iv. Hauniae [Copenhague]: Libraria Gyldendaliana.

Brahe, T. (1596). *Epistolarum astronomicarum liber primus*, Uraniburg. In Tycho Brahe, *Opera omnia*, ed. by J.L.E. Dreyer, vol. vi. Hauniae [Copenhague]: Libraria Gyldendaliana.

Brahe, T. (1602). *Astronomiae instauratae progymnasmata*, Uraniburg and Prague. In Tycho Brahe, *Opera omnia*, ed. by J.L.E. Dreyer, vol. ii–iii. Hauniae [Copenhague]: Libraria Gyldendaliana.

Bruno, G. (2006^2). *De l'infinito, universo e mondi / De l'infini, de l'univers et des mondes*. In Giordano Bruno, *Opere complete / Oeuvres complètes*, ed. by G. Aquilecchia a. o., vol. iv. Paris: Les Belles Lettres.

Bubenheimer, U. (1995). Wilhelm Schickard in Kontext einer religiösen Subkultur. In F. Seck (Ed.), *Zum 400. Geburtstag von Wilhelm Schickard: Zweites Tübinger Schickard-Symposion 25. bis 27. Juni 1992*, pp. 67–92. Sigmaringen: Thorbecke.

Bucciantini, M. (2003). *Galileo e Keplero. Filosofía, cosmologia e teologia nell'età della Controriforma*. Turin: Enaudi.

Bucciantini, M.—Camerota, M.—Giudice, F. (2012). *Il telescopio di Galileo: Una storia europea*. Turin: Einaudi.

Burmeister, K.H. (2015). *Magister Rheticus und seine Schulgesellen: Das Ringen um Kenntnis und Durchsetzung des heliozentrischen Weltsystems des Kopernikus um 1540/50*. Konstance and Munich: UVK Verlagsgesellschaft.

Busch, G. (1577). *Beschreibung von zugehörigen Eigenschafften und natürlichen Influentz des grossen und erschrecklichen Cometen welcher in diesem 1577. Jahre erschienen*. Erfurt: Esaias Mechlern.

Cardano, G. (1554). *Claudii Ptolemaei Pelusiensis libri quatuor de Astrorum Iudiciis cum expositione Hieronymi Cardani*. Basel: Henricus Petri.

Cardano, G. (2004). *De subtilitate*, critical edition by Elio Nenci, vol I, books I–VII. Milan: Franco Angeli.

Carolino, L.M. (2012). Manuel Bocarro Francês, the Comet of 1618, and the Impact of Stoic Cosmology in Portugal. In M.Á. Granada (Ed.), *Novas y cometas entre 1572 y 1618: Revolución cosmológica y renovación política y religiosa*, pp. 195–224. Barcelona: Edicions de la Universitat de Barcelona.

Cedillo Díaz, J. (2019). *Ydea astronomica de la fabrica del mundo y movimiento de los cuerpos celestiales. Traducción de 'De revolutionibus i–iii', de Nicolás Copérnico*, edited by Miguel Á. Granada and Félix Gómez Crespo. Barcelona: Edicions de la Universitat de Barcelona.

Christianson, J.R. (1979). Tycho Brahe's German Treatise on the Comet of 1577: A Study in Science and Politics. *Isis* 70, 110–140.

Christianson, J.R. (2020). *Tycho Brahe and the Measure of the Heavens*. London: Reaktion.

Cicero (1933). *De natura deorum. Academica*, translated by H. Rackham, (Loeb Classical Library, no. 268). Cambridge MA—London: Harvard University Press—William Heinemann.

Clavius, C. (1583). *Epitome arithmeticae practicae*. Rome: Dominicus Basa.

Copernic, N. (2015). *De revolutionibus orbium coelestium / Des révolutions des orbes célestes*, édition critique, traduction et notes par M.–P. Lerner, A.–P. Segonds et J.–P. Verdet, 3 vols. Paris: Les Belles Lettres.

Copernicus, N. (1543). *De revolutionibus orbium coelestium libri sex*. Nuremberg: Apud Joh. Petreium.

Copernicus, N. (1992). *On the Revolutions*, translation and commentary by E. Rosen. Baltimore—London: The Johns Hopkins University Press.

Crüger, P. (1619). *Uranodromus cometicus. Ein ausführlicher Tractat vom grossen Cometen dess 1618 Jahrs*. Danzig: Andreas Hünefeld.

Cysat, J.B. (1619). *Mathemata astronomica de loco, motu, magnitudine et causis cometae qui sub finem Anni 1618. et initium Anni 1619. in coelo fulsit*. Ingolstadt: Eder.

Dall'Olmo, U. (1980). Latin Terminology Relating to Aurorae, Comets, Meteors and Novae. *Journal for the History of Astronomy* 11, 10–27.

Diodorus Siculus (1933–1967). [*The Library of History*] *in Twelve Volumes*, translated by

C.H. Oldfather, C.L. Sherman and others, (Loeb Classical Library, nos. 279, 303, 340, 375, 377, 384, 389–390, 399, 409, 422–423). Cambridge MA – London: Harvard University Press—William Heinemann.

Drake, S. and O'Malley, C.D. (1960). *The Controversy of the Comets of 1618*. Philadelphia: University of Pennsylvania.

Einhard (1880). *The Life of Charlemagne*, translated by S.E. Turner. New York: Harper & Brothers.

Fabbri, N. (2012). The Moon as Another Earth: What Galileo Owes to Plutarch, *Galilaeana*, IX, 103–135.

Fabricius, J. (1611). *De maculis in sole observatis et apparente earum cum Sole conversione, narratio*. Wittenberg: Laurentius Seuberlichius.

Faulhaber, J. (1604). *Arithmetischer Cubicossischer Lustgarten*. Tübingen: bey Erhardo Cellio.

Faulhaber, J. (1612). *Newer Mathematischer Kunstspiegel. Darinnen fürnemblich dreyerley Stuck zu sehen. Als erstlich ein gründtliche Verzeichnuss der wunderbarlichen Natur und Eigenschafften etlicher Zahlen Danielis und der Offenbarung Sanct Johannis*. Ulm: Johann Meder.

Faulhaber, J. (1613). *Andeutung einer unerhörten newen Wunderkunst. Welche der Geist Gottes in etlichen Prophetischen und Biblischen Geheimnuss Zahlen, biss auff die letzte Zeit hatt wöllen versigelt und verborgen halten*. Nuremberg: Abraham Wagenmann.

Faulhaber (1613b). *Ansa inauditae et admirabilis novae artis, quam Spiritus Dei arcanis aliquot propheticis et biblicis numeris ad ultima haec tempora obsignare et operire voluit*. Frankfurt: Wagemann.

Faulhaber (1613c). *Himlische gehaime Magia oder Newe cabalistische Kunst und Wunderrechnung vom Gog und Magog*. Nuremberg: Abraham Wagenmann.

Faulhaber, J. (1618–1619). *Fama syderea nova. Gemein offentliches Aussschreiben* [...] *Anlangend den Neuen: und durch ein sonderbare Invention lang zuvor prognosticirten Cometstern so den 6. Monatstag Decembr. dess ablaufenden 1618. Jahrs, an alle Philosophos, Mathematicos, Sonderlich Astronomos und Gelehrte dess gantzen Deutschlands*. Nuremberg: J.G. Goldtbeeg.

Frank, G. (1997). Veit Amerbach (1503–1557). Von Wittenberg nach Ingolstadt. In Heinz Scheible (Ed.), *Melanchthon in seinen Schülern*, pp. 103–128. Wiesbaden: Harrassowitz.

Freedman, J.S. (1997). The Career and Writings of Bartholomew Keckermann (d. 1609). *Proceedings of the American Philosophical Society* 141, 305–364.

Frisius, G. (1545). *De radio astronomico et geographico liber*. Antwerp: G. Bontius.

Funkenstein, A. (1986). *Theology and the Scientific Imagination: From the Middle Ages to the Seventeenth Century*. Princeton, NJ: Princeton University Press.

Gäbe, L. (1972). *Descartes' Selbstkritik. Untersuchungen zur Philosophie des jungen Descartes*. Hamburg: Felix Meiner.

Galilei, G. (1989). *Sidereus Nuncius or The Sidereal Messenger*, translated with introduction, conclusion and notes by A. Van Helden. Chicago and London: Chicago University Press.

Galilei, G.—Scheiner, C. (2010). *On Sunspots*, translated and introduced by E. Reeves and A. Van Helden. Chicago and London: Chicago University Press.

Galileo—Grassi, (2016). *Cometas, ciencia y religión. La polémica Galileo—Grassi*, Introduction and notes by A. Beltrán Marí, translated by A. Beltrán Marí and E. Arrigas Álvarez. Madrid: Tecnos.

Garcaeus, J. (1568). *Meteorologia. Addita sunt tabellae, quae totam meteororum doctrinam complectuntur, et exempla sacra et prophana, multorum seculorum, quibus haec materia scholasticorum causa illustrata est*. Wittenberg: Iohan Schwertel.

Gassendi, P. (1630). *Epistolica exercitatio in qua, principia philosophiae Roberti Fluddi medici reteguntur* [...]. *Cum appendice aliquot observationum coelestium*. Paris: Sebastianum Cramoisy.

Gemma, C. (1578). *De prodigiosa specie naturaque cometae qui nobis effulsit altior Lunae sedibus* [...] *anno 1577*. Antwerp: Christophorus Plantinus.

Gilly, C. (2002). Der 'Löwe von Mitternacht', der 'Adler' und der 'Endchrist'. Die politische, religiöse und chiliastische Publizistik in den Flugschriften, illustrierten Flugblättern und Volksliedern des Dreissigjährigen Krieges. In *Rosenkreuz als europäisches Phänomen im 17. Jahrhundert*, pp. 249–268. Amsterdam: In de Pelikaan.

Gilly, C. (2012). Las novas de 1572 y 1604 en los manifiestos rosacruces y en la literatura teosófica y escatológica alemana anterior a la Guerra de los Treinta Años. In M.Á. Granada (Ed.), *Novas y cometas entre 1572 y 1618: Revolución cosmológica y renovación política y religiosa*, pp. 275–331. Barcelona: Edicions de la Universitat de Barcelona.

Gindhart, M. (2006). *Das Kometenjahr 1618: Antikes und zeitgenössisches Wissen in der frühneuzeitlichen Kometenliteratur des deutschsprachigen Raumes*. Wiesbaden: Dr. L. Reichert Verlag.

Gingerich, O. (2002). *An Annotated Census of Copernicus'* De revolutionibus (*Nuremberg, 1543 and Basel, 1566*). Leiden–Boston–Köln: Brill.

Gliozzi, G. (1977). *Adamo e il nuovo mondo*. Florence: La Nuova Italia

Gloriosi, G.C. (1624). *De cometis dissertatio astronomica publice habita in gymnasio patavino anno Domini* M DC XIX. Venice: Typographia Variscina.

Goldstein, B.R. (1965). Evidence for a Supernova of A.D. 1006, *The Astronomical Journal* 70, 105–114.

Granada, M.Á. (2002). *Sfere solide e cielo fluido: Momenti del dibattito cosmologico nella seconda metà del Cinquecento*. Milan: Guerini e Associati.

Granada, M.Á. (2006). Did Tycho Eliminate the Celestial Spheres before 1586?, *Journal for the History of Astronomy* 37, 125–145.

Granada, M.Á. (2007). Michael Maestlin and the New Star of 1572. *Journal for the History of Astronomy* 38, 99–124.

Granada, M.Á. (2009). La théorie des comètes de Helisaeus Roeslin. In M.A. Granada and E. Mehl (Eds.), *Nouveau ciel, nouvelle terre: La révolution copernicienne dans l'Allemagne de la Réforme (1530–1630)*, 207–244. Paris: Les Belles Lettres.

Granada, M.Á. (2013). Tycho Brahe's Anti-Copernican Campaign: His Criticism of Maestlin and Thomas Digges in the *Astronomiae instauratae progymnasmata*. In D. Tessicini—Patrick J. Boner (Eds.), *Celestial Novelties on the Eve of the Scientific Revolution, 1540–1630*, pp. 185–207. Florence: Olschki.

Granada, M.Á. (2014). Michael Maestlin and his Unpublished Treatise on the Nova of 1604. *Journal for the History of Astronomy* 45, 91–122 (the edition of the Latin text is accompanied by an English translation by P.J. Boner).

Granada, M.Á. (2016). Michael Maestlin and the Comet of 1618. In M.Á. Granada, P. J. Boner and D. Tessicini (Eds.), *Unifying Heaven and Earth: Studies in the History of Early Modern Cosmology*, pp. 239–290. Barcelona: Edicions de la Universitat de Barcelona.

Granada, M.Á. (2017). Helisaeus Röslin: 4 Esdras et l'avènement du royaume du Christ. In É. Mehl (Ed.). *Le Temps des Astronomes: L'astronomie et le décompte du temps de Pierre d'Ailly à Newton*, pp. 123–149.

Granada, M.Á. (2019). Bartholomaeus Keckermann and Christoph Hunichius on Novas and Comets at the Beginning of the 17th Century: Two Opposing Views on the Relation between Natural Philosophy and Mathematics. In P.D. Omodeo and V. Wels (Eds.), *Natural Knowledge and Aristotelianism at Early Modern Protestant Universities*, pp. 209–234. Wiesbaden: Harrassowitz.

Granada, M.Á., Mosley, A. and Jardine, N. (2014). *Christoph Rothmann's Discourse on the Comet of 1585: An Edition and Translation with Accompanying Essays*. Leiden: Brill.

Guicciardini, L. (1580). *Niderlands Beschreibung: In welcher aller darinn begriffnen Landtschafften, Fürstenthumben* [...] *Ursprung und Auffgang eigentlich erklärt wird*, transl. by D. Federman von Memmingen. Basel: Sebastian Henricpetri.

Habrecht, I. (1618). *Kurtze und gründliche Beschreibung eines Newen ungewohnlichen Sterns oder Cometen* [...] *im November und December diß 1618. Jahr erschienen*. Strasbourg: bey Johann Carolo.

Hagecius, T. (1574). *Dialexis de novae et prius incognitae stellae inusitatae magnitudinis & splendidissimi luminis apparitione et de eiusdem stellae vero loco constituendo*. Frankfurt: n.p.

Hagecius, T. (1578). *Descriptio cometae qui apparuit Anno Domini MDLXXVII à IX die Novembris usque ad XIII. diem Ianuarii, Anni &c. LXXVIII*. Prague: Georgius Melantrichus.

Hallyn, F. (2008). *Gemma Frisius, arpenteur de la terre et du ciel*. Paris: Honoré Champion.

Hamel, J. (2016). Simon Marius, seine Arbeiten über die Kometen von 1596 und 1618 im Zusammenhang mit der Kometenforschung seiner Zeit. In H. Gaab—P. Leich

(Eds.), *Simon Marius und seine Forschung*, pp. 199–216, (Acta Historica Astronomiae 57). Leipzig: Akademische Verlagsanstalt.

Hamel, J. (2017). Landgraf Wilhelm IV und Michael Mästlin in der Polemik um die Gregorianische Kalenderreform. In É. Mehl (Ed.), *Le Temps des Astronomes: L'astronomie et le décompte du temps de Pierre d'Ailly à Newton*, pp. 231–268. Paris: Les Belles Lettres.

Hamesse, J. (1974). Les *Auctoritates Aristotelis: Un florilège médiéval. Étude historique et édition critique*. Louvain–Paris: Publications Universitaires and Béatrice Nauwelaerts.

Hamilton, A. (1999). *The Apocryphal Apocalypse: The Reception of the Second Book of Esdras (4 Ezra) from the Renaissance to the Enlightenment*. Oxford: Clarendon Press.

Hawlitschek, K. (1995). *Johann Faulhaber 1580–1635: Eine Blütezeit der mathematischen Wissenschaften in Ulm*. Ulm: Stadtbibliothek Ulm.

Hawlitschek, K. (2006). *Johann Faulhaber und René Descartes auf dem Weg zur modernen Wissenschaft. Ulm 1619*. Ulm: Stadtbibliothek Ulm.

Heath, T. (1981). *Aristarchus of Samos: The Ancient Copernicus*. New York: Dover.

Hebenstreit, J.B. (1618). *Cometen Fragstuck auß der reinen Philosophia, Bey Anschauung deß in diesem 1618. Jahr in dem Obern Lufft schwebenden Cometen erläutert*. Ulm: Johann Meder.

Heidarzadeh, T. (2008). *A History of Physical Theory of Comets, From Aristotle to Whipple*, (Archimedes: New Studies in the History and Philosophy of Science and Technology, 19). New York: Springer.

Hellman, C.D. (1944). *The Comet of 1577. Its Place in the History of Astronomy*. New York: Columbia University Press.

Herodotus (1920–1925). [*The Histories*], translated by A.D. Godley, (Loeb Classical Library, nos. 117–120). London-Cambridge MA: Heinemann-Harvard University Press.

Jaki, S. (1973). *The Milky Way: An Elusive Road for Science*. Newton Abbot: David Charles.

Jarrell, R.A. (1971). *The Life and Scientific Work of the Tübingen Astronomer Michael Maestlin, 1550–1631*. Ph.D. Thesis, University of Toronto [A French translation by Nicolas Roudet is currently in preparation: Michael Maestlin (1550-1631). Paris: Les Belles Lettres].

Jarrell, R.A. (1975). Mästlin's Place in Astronomy. *Physis* 17, 5–20.

Jarrell, R.A. (1981). Astronomy at the University of Tübingen: The Work of Michael Maestlin. In F. Seck (Ed.), *Wissenschaftsgeschichte um Wilhelm Schickard*, pp. 9–19. Tübingen: J.C.B. Mohr.

Jensen, D. (2006). *The Science of Stars in Danzig from Rheticus to Hevelius*. Ph.D. Thesis, University of California, San Diego. UMI Dissertation Services.

Jerratsch, A. (2020). *Der frühneuzeitliche Kometendiskurs im Spiegel deutschsprachiger Flugschriften*. Stuttgart: Franz Steiner.

Jervis, J.L. (1985). *Cometary Theory in Fifteenth-Century Europe*. Dordrecht–Boston–Lancaster: Reidel.

Kapoor, R.C. (2016). Nūr ud-Dīn Jahāngīr and Father Kirwitzer: The Independent Discovery of the Great Comets of November 1618 and the First Astronomical Use of the Telescope in India. *Journal of Astronomical History and Heritage* 19, 264–297.

Keckermann, B. (1606). *Disputationes philosophicae, Physicae praesertim, quae in Gymnasio Dantiscano ad lectionum philosophicarum cursum paulo plus biennio publice institutae et habitae sunt sub praesidio Bartholomaei Keckermanni*. Hanau: Apud Guilielmum Antonium.

Keckermann, B. (1623). *Systema physicum septem libris adornatum, et Anno Christi MDCVII publice propositum in Gymnasio Dantiscano*. Hanau: Apud Guilielmum Antonium.

Kepler, J. (1596). *Mysterium cosmographicum*. In Max Caspar at al. (Eds.), *J. Kepler, Gesammelte Werke*, vol. i, pp. 1–145. Munich: Beck.

Kepler, J. (1606). *De stella nova in pede Serpentarii*. In Max Caspar et al. (Eds.), *J. Kepler, Gesammelte Werke*, vol. i, pp. 151–390.

Kepler, J. (1616). *Ephemerides novae motuum coelestium, ab anno vulgaris aerae MDCXVII*. In Max Caspar et al. (Eds.), *J. Kepler, Gesammelte Werke*, vol. xi, 1, pp. 7–135. Munich: Beck.

Kepler, J. (1618). *Prognosticum Astrologicum auff das Jahr MDCXVIII*. In Max Caspar et al. (Eds.), *J. Kepler, Gesammelte Werke*, vol. xi, 2, pp. 154–172. Munich: Beck

Kepler, J. (1619). *De cometis libelli tres*. In Max Caspar et al. (Eds.), *J. Kepler, Gesammelte Werke*, vol. viii, pp. 132–262. Munich: Beck.

Kepler, J. (1619a). *Prognosticum auff das Jahr 1620*. In Max Caspar et al. (Eds.), *J. Kepler, Gesammelte Werke*, vol. xi, 2, pp. 189–215. Munich: Beck.

Kepler, J. (1621). *Mysterium Cosmographicum* [editio altera]. In Max Caspar et al. (Eds.), *J. Kepler, Gesammelte Werke*, vol. viii, pp. 5–128. Munich: Beck.

Kepler, J. (1965). *Kepler's Conversation with Galileo's Sidereal Messenger*, translated by E. Rosen. Madison: University of Wisconsin Press.

Kokott, W. (1994). *Die Kometen der Jahre 1531 bis 1539 und ihre Bedeutung für die spätere Entwicklung der Kometenforschung*. Stuttgart: GNT–Verlag.

Kronk, G.W. (1999). *Cometography: A Catalogue of Comets. Volume 1: Ancient—1799*. Cambridge: Cambridge University Press.

Kühlmann, W. (1995). Wilhelm Schickard—Wissenschaft und Reformbegehren in der Zeit des Konfessionalismus. In F. Seck (Ed.), *Zum 400. Geburtstag von Wilhelm Schickard: Zweites Tübinger Schickard-Symposion 25. bis 27. Juni 1992*, pp. 41–66. Sigmaringen: Thorbecke.

Lerner, M.-P. (1995). L'entrée de Tycho Brahe chez les jésuites ou le chant du cygne de Clavius. In L. Giard (Ed.), *Les jésuites à la Renaissance: Système éducatif et production du savoir*, pp. 145–185. Paris: P.U.F.

Lerner, M.-P. (2008). *Le monde des sphères*, 2 vols. Paris: Les Belles Lettres.

Macrobius (2001–2003). *Commentaire au songe de Scipion, livres I–II*, edited and translated by M. Armissen-Marchetti. Paris: Les Belles Lettres.

Maestlin, M. (1573). *Demonstratio astronomica loci stellae novae, tum respectu centri mundi, tum respectu signifieri et aequinoctialis*. In N. Frischlin, *Consideratio novae stellae, quae mense novembri, Anno Salutis M.D.LXXII. in signo Cassiopeae apparuit*, pp. 27–32. Tübingen: Georg Gruppenbach.

Maestlin, M. (1578). *Observatio et demonstratio cometae aetherei, qui anno 1577. et 1578. constitutus in sphaera Veneris, apparuit, cum admirandis eius passionibus, varietate scilicet motus, loco, orbe, distantia a terrae centro, etc adhibitis demonstrationibus geometricis et calculo arithmetico, cuiusmodi de alio quoquam cometa nunquam visa est*. Tübingen: Georg Gruppenbach.

Maestlin, M. (1581). *Consideratio et observatio cometae aetherei astronomica, qui anno MDLXXX. mensibus octobri, novembri et decembri, in alto aethere apparuit*. Heidelberg: Iacobus Mylius.

Maestlin, M. (1606). *Disputatio de multivariis motuum planetarum in coelo apparentibus irregularitatibus*. Tübingen: Ex officina Gruppenbachiana.

Maestlin, M. (1610). *Epitome astronomiae*. Tübingen: Philippus Gruppenbach.

Martin, C. (2011). *Renaissance Meteorology: Pomponazzi to Descartes*. Baltimore: Johns Hopkins University Press.

Mauder, H. (1995). Die Kometenschriften von Schickard und Kepler. In Friedrich Seck (Ed.), *Zum 400. Geburtstag von Wilhelm Schickard. Zweites Tübinger Schickard-Symposion 25. bis 27. Juni 1992*, pp. 151–166. Sigmaringen: Thorbecke.

Mehl, É. (2013). Comètes et taches solaires en Allemagne (1610–1630): L'aile hétérodoxe (Faulhaber, Mayr, Mögling) et le point de départ de la 'fable du monde' cartésienne. In D. Tessicini—Patrick J. Boner (Eds.), *Celestial Novelties on the Eve of the Scientific Revolution, 1540–1630*, pp. 231–255. Florence: Olschki.

Mehl, É. (2019). *Descartes en Allemagne, 1619–1620: Le contexte allemand de l'elaboration de la science cartésienne*. Paris: Press Universitaires de Strasbourg.

Melanchthon, P. (1595). *Liber de anima, recognitus ab autore*. Wittenberg: Georgius Mullerus.

Melanchthon, P. (1999). *Orations on Philosophy and Education*, ed. by S. Kusukawa, trans. by C.F. Salazar. Cambridge: Cambridge University Press.

Methuen, C. (1996). Maestlin's Teaching of Copernicus: The Evidence of His University Textbook and Disputations. *Isis* 87, 230–247.

Methuen, C. (1998). *Kepler's Tübingen: Stimulus to a Theological Mathematics*. Aldershot: Ashgate.

Mizaldus, A. (1549). *Cometographia: crinitarum stellarum quas mundus nunquam impune vidit, aliorumque ignitorum aëris Phaenomenων, naturam & portenta duobus libris philosophicè iuxtà ac astronomicè expediens*. Paris: Christianus Wechelus.

Müller, P. (1619). *De cometa anni M.DC.XVIII. commentatio physicomathematica specialis & generalis*. Leipzig: Typis Grossianis.

Navarro Brotóns, V. (2012). Las novedades celestes en España entre 1572 y 1618. In M.Á. Granada (Ed.), *Novas y cometas entre 1572 y 1618: Revolución cosmológica y renovación política y religiosa*, pp. 15–41. Barcelona: Edicions de la Universitat de Barcelona.

Neumann, U. (1995). 'Olim, da die Rosen Creutzerey noch florirt, Theophilus Schweighardt genant': Wilhelm Schickards Freund und Briefpartner Daniel Mögling (1596–1635). In F. Seck (Ed.), *Zum 400. Geburtstag von Wilhelm Schickard: Zweites Tübinger Schickard-Symposion 25. bis 27. Juni 1992*, pp. 93–115. Sigmaringen: Thorbecke.

Nolthius, A. (1578). *Observatio und Beschreibung des Cometen, welcher im Novembri und Decembri des 77. und noch im Januario dieses 78. Jharsz erschienen*. Erfurt: G. Bawman.

Nonius, P. (1542). *De crepusculis liber unus. Item Allacen Arabis vetustissimi, de causis crepusculorum liber unus*. Lisbon: Ludovicus Rodericus.

Osiander, L. (1619). *Admonitio de quorundam, ad praesentia haec periculosa tempora spectantium vaticiniorum, per fanaticos quosdam, novorum motuum fautores, in publicum sparsis corruptelis*. Tübingen: Apud Eberhardum Wildium.

Ovid (1984). *Metamorphoses*, translated by F.J. Miller (revised by G.P. Goold), (Loeb Classical Library, nos. 42–43). Cambridge, MA and London: Harvard University Press–William Heinemann.

Pausanias (1918–1935). *Description of Greece*, 5 vols., English translation by W.H.S. Jones, (Loeb Classical Library, nos. 93, 188, 272, 297–298). Cambridge, MA and London: Harvard University Press–William Heinemann.

Pena, J. (1557). *Euclidis Optica et Catoptrica, nunquam antehac Graece aedita. Eadem Latine reddita per Ioannem Penam. His praeposita est eiusdem Ioannis Penae de usu optices praefatio*. Paris: Wechel.

Pingré, A.G. (1783–1784). *Cométographie ou traité historique et théorique des comètes*. Paris: Imprimerie Royale.

Plinius, C. (1601–1616). *Historiae mundi libri XXXVII*, 3 vols., edited by S. Gelen. S.l.: Jacobus Stoer.

Pliny (1938–1952). *Natural History in Ten Volumes*, English Translation, by H. Rackham, (Loeb Classical Library, nos. 330, 352–353, 370–371, 392–394, 418–419). Cambridge, MA–London: Harvard University Press and William Heinemann.

Plutarch (1516). *De philosophorum placitis*, trans. by G. Budé. Strasbourg: Mathias Schürer.

Plutarch (1957). *Concerning the Face which Appears in the Orb of the Moon* (*De facie quae in orbe Lunae apparet*), translated and commented by H. Cherniss, (Loeb Classical Library, no. 406). Cambridge, MA–London: Harvard University Press and William Heinemann.

Ptolemy (1584). *Geographiae Libri Octo, recogniti iam et diligenter emendati cum tabulis geographicis ad mentem auctoris restitutis ac emendatis, per Gerardum Mercatorem.* Cologne: Gottfried von Kempen.

Ptolemy (1984). *Almagest*, translated and annotated by G.J. Toomer. London: Duckworth.

Randles, W.G.L. (2000). *Geography, Cartography and Nautical Science in the Renaissance: The Impact of the Great Discoveries.* (Variorum Collected Studies) Aldershot: Ashgate.

Regiomontanus, J.M. (1531). *De Cometae magnitudine, longitudine ac de loco eius vero, problemata XVI.* Nuremberg: Apud Fridericum Peypus.

Reinhold, E. (1551). *Tabulae prutenicae.* Tübingen: Per Ulricum Morhardum.

Remus Quietanus, J. (1619). *Observationes et descriptiones duorum cometarum qui anno 1618 mense novembri usque ad finem anni currentis in aetherea regione visi sunt.* Innsbruck: Apud Danielem Paur.

Rheticus, G.J. (1982). *Narratio prima.* Édition critique, traduction française et commentaire par H. Hugonnard–Roche et J.–P. Verdet avec la collaboration de M.–P. Lerner et A. Segonds. Wroclaw–Warsaw–Krakow–Gdansk–Lodz: Ossolineum

Risnerus, F. (1572). *Opticae Thesaurus Alhazeni Arabis libri septem, nunc primum editi. Eiusdem liber de crepusculis et nubium accensionibus. Item Vitellionis Thuringopoloni libri X*, ed. by F. Risnerus. Basel: Per Episcopios.

Rockenbach, A. (1602). *De cometis, tractatus novus methodicus.* Wittenberg: Ex officina Cratoniana.

Roeslin, H. (1597). *Tractatus meteorastrologiphysicus.* Strasbourg: B. Jobins Erben.

Rosselló Botey, V. (2000). *Tradició i canvi científic en l'astronomia espnyola del segle XVII.* Valencia: Universitat de València.

Röttel, K., Ed. (1995). *Peter Apian: Astronomie, Kosmographie und Mathematik am Beginn der Neuzeit, mit Ausstellungskatalog.* Buxheim-Eichstätt: Polygon–Verlag.

Ruffner, J.A. (1971). The Curved and the Straight: Cometary Theory from Kepler to Hevelius. *Journal for the History of Astronomy* 2, 178–194.

Sabra, A.I. (1967). The Authorship of the *Liber de crepusculis. Isis* 58, 77–85.

Sabra, A.I. (1994). The Andalusian Revolt against Ptolemaic Astronomy: Averroes and al-Biṭrūjī. In A.I. Sabra, *Optics, Astronomy and Logic in Arabic Science and Philosophy*, pp. 133–153. Ashgate: Variorum Reprints.

Schickard, W. (1619). *Cometen Beschreibung In zwen underschidliche Partes abgetheilt, deren Erster Von denzselbigen ins gemein: der Ander Von allen Insonderheit, sonderlich aber denen drey Jüngsten, Im abgeloffnen 1618 Jahr erschienen, aussführlich handelt.* WLB, Cod. math. 4. 43.

Schneider, I. (1993). *Johannes Faulhaber 1580–1635: Rechenmeister in einer Welt des Umbruchs.* Basel–Boston–Berlin: Birkhäuser.

Schneider, I. (2004). Between Rosicrucians and Cabbala: Johannes Faulhaber's Math-

ematics of Biblical Numbers. In T. Koetsier–L. Bergmans (Eds.), *Mathematics and the Divine: A Historical Study*, pp. 311–330. Amsterdam: Elsevier.

Schramm, M. (1978). Der Astronom. In F. Seck (Ed.), *Wilhelm Schickard 1592–1635: Astronom, Geograph, Orientalist, Erfinder der Rechenmaschine*, pp. 129–287. Tübingen: J.C.B. Mohr (Paul Siebeck).

Schweighardt, T. [= Daniel Mögling] (1618). *Speculum Sophicum Rhodo-Stauroticum. Das ist: Weitläuffige Entdeckung deß Collegii unnd axiomatum von der sondern erleuchten Fraternitet Christ-RosenCreutz: allen der wahren Weißheit Begirigen Expectantes zu fernerer Nachrichtung, den unverständigen Zoilis aber zur ußlöschlicher Schandt und Spott.* s.l. [Frankfurt: Joh. Theodor de Bry].

Scultetus, B. (1578). *Cometae, anno humanitatis MDLXXVII* [...] *in sublunari regione adparentis, descriptio*. Görlitz: Ambrosius Fritsch.

Seck, F. (2002). Der Briefwechsel zwischen Kepler und Mästlin. In Gerhard Betsch and Jürgen Hamel (Eds.), *Zwischen Copernicus und Kepler—M. Michael Maestlinus Mathematicus Goeppingensis 1550–1631*, (Acta Historica Astronomiae 17), pp. 110–121. Frankfurt: Harri Deutsch.

Seneca (1972). *Naturales quaestiones*, with an English translation by Thomas H. Corcoran, (Loeb Classical Library, nos. 450, 457) London–Cambridge, MA: William Heinemann Ltd and Harvard University Press.

Shalev, Z. (2003). Sacred Geography, Antiquarianism and Visual Erudition: Benito Arias Montano and the Maps in the Antwerp Polyglott Bible. *Imago Mundi* 55, 56–80.

Shea, W.R. (1970). Galileo, Scheiner, and the Interpretation of Sunspots. *Isis* 61, 498–519.

Siebert, H. (2006). *Die große kosmologische Kontroverse: Rekonstruktionsversuche anhand des Itinerarium exstaticum von Athanasius Kircher SJ (1602–1680)*. Stuttgart: Franz Steiner.

Snel, W. (1619). *Descriptio cometae qui anno 1618, mense novembri primum effulsit*. Leiden: Elzevier,

Stephenson, F.R., Clark, D.H., Crawford, D.F. (1977). The Supernova of AD 1006. *Monthly Notices of the Royal Astronomical Society* 180, 567–584.

Strabo (1917–1932). *Geography*, with an English translation by H.L. Jones (Loeb Classical Library, nos. 49–50, 182, 196, 211, 223, 241, 267). London–Cambridge, MA: William Heinemann Ltd and Harvard University Press.

Telesio, B. (2012). *Sobre los cometas y la Vía Láctea*, edition of the Latin text and translation by M.Á. Granada. Madrid: Tecnos.

Tessicini, D. (2010). Comets. In E. Canone and G. Ernst (Eds.), *Enciclopedia Bruniana e Campanelliana*, vol II, cols 34–43. Pisa-Rome: Fabrizio Serra.

Urbanek, V. (2002). The Comet of 1618: Eschatological Expectations and Political Prognostications during the Bohemian Revolt. In J.R. Christianson, A. Hadravová, P. Hadrava, M. Solc (Eds.), *Tycho Brahe and Prague*. (Acta Historica Astronomiae 16), pp. 282–291. Frankfurt: Harri Deutsch.

Van Dülmen, R. (1971). Ein Gutachten W. Schickards über Rosenkreutzerschriften. *Zeitschrift für Religions- und Geistesgeschichte* 23, 254–259.

Van Helden, A. (1985). *Measuring the Universe: Cosmic Dimensions from Aristarchus to Harvey*. Chicago and London: University of Chicago Press.

Van Nouhuys, T. (1998). *The Age of Two-faced Janus: The Comets of 1577 and 1618 and the Decline of the Aristotelian World View in the Netherlands*. Leiden– Boston–Köln: Brill.

Vermij, R. (2010). A Science of Signs: Aristotelian Meteorology in Reformation Germany. *Early Science and Medicine* 15, 648–674.

Vögelin, J. (1533). *Significatio cometae qui anno MDXXXII. apparuit cum passionibus eius scitu iucundissimis*. Vienna: In Hagecius (1574, 150–167; with 1527 as the wrong date of the comet).

Westman, R.S. (1972). The Comet and the Cosmos: Kepler, Maestlin and the Copernican Hypothesis. *Studia Copernicana* 5, 7–30.

Westman, R.S. (2011). *The Copernican Question: Prognostication, Skepticism, and Celestial Order*. Berkeley-Los Angeles-London: University of California Press.

Index of Biblical Passages

Old Testament

New Testament

Index of Persons

Pre-1900

Post-1900

Printed in the United States
by Baker & Taylor Publisher Services